TRILLIONS FOR MILITARY TECHNOLOGY

TRILLIONS FOR MILITARY TECHNOLOGY

HOW THE PENTAGON INNOVATES AND WHY IT COSTS SO MUCH

John A. Alic

First published in 2007 by
PALGRAVE MACMILLAN™
175 Fifth Avenue, New York, N.Y. 10010 and
Houndmills, Basingstoke, Hampshire, England RG21 6XS
Companies and representatives throughout the world.

PALGRAVE MACMILLAN is the global academic imprint of the Palgrave Macmillan division of St. Martin's Press, LLC and of Palgrave Macmillan Ltd. Macmillan® is a registered trademark in the United States, United Kingdom and other countries. Palgrave is a registered trademark in the European Union and other countries.

ISBN-13: 978–1–4039–8426–5
ISBN-10: 1–4039–8426–3

Library of Congress Cataloging-in-Publication Data

Alic, John A.
    Trillions for military technology : how the Pentagon innovates and why it costs so much / by John A. Alic.
        p. cm.
    Includes bibliographical references and index.
    ISBN 1–4039–8426–3 (alk. paper)
        1. Military research—United States—Costs. 2. United States—Armed Forces—Weapons systems—Technological innovations—Costs. 3. United States. Dept. of Defense—Procurement. 4. Military weapons—Technological innovations—United States—Costs. I. Title. II. Title: How the Pentagon innovates and why it costs so much.

U393.5.A19 2007
355'.070973—dc22   2006052520

A catalogue record for this book is available from the British Library.

Design by Newgen Imaging Systems (P) Ltd., Chennai, India.

First edition: September 2007

10   9   8   7   6   5   4   3   2   1

Printed in the United States of America.

For Annie and Zoë-Maud and Mac

CONTENTS

LIST OF FIGURES

LIST OF TABLES

SELECTED ABBREVIATIONS

AEC	Atomic Energy Commission
AFOSR	Air Force Office of Scientific Research
CBO	Congressional Budget Office
D&D	design and development
DARPA	Defense Advanced Research Projects Agency
DOD	Department of Defense
DSB	Defense Science Board
FFRDC	federally funded research and development center
GAO	General Accounting Office; since 2004 Government Accountability Office
GPS	Global Positioning System
ICBM	intercontinental ballistic missile
IC	integrated circuit
JCS	Joint Chiefs of Staff
NACA	National Advisory Committee for Aeronautics
NASA	National Aeronautics and Space Administration
NATO	North Atlantic Treaty Organization
NRL	Naval Research Laboratory
NSF	National Science Foundation
OMB	Office of Management and Budget
ONR	Office of Naval Research
OSD	Office of the Secretary of Defense
OSRD	Office of Scientific Research and Development
QDR	Quadrennial Defense Review
RDT&E	research, development, test, and evaluation
RMA	Revolution in Military Affairs
SAC	Strategic Air Command
UAV	unmanned aerial vehicle

CHAPTER 1

CHOOSING WEAPONS

Since the close of World War II, the United States has spent some $1.3 trillion on military R&D (equivalent to $2 trillion in year 2000 dollars). Procurement outlays—expenditures for equipment and systems based on that R&D—added another $2.3 trillion (about $3.4 trillion in 2000 dollars). Defense acquisition—the term encompasses both R&D and procurement—has thus consumed about $3.6 trillion, more than one-third of postwar U.S. defense spending. And over the first decade of the twenty-first century, the United States will spend nearly $1.5 trillion on new weapons systems.[1]

This book is concerned with where that money goes. As we will see, the Korean War was the primary impetus behind the U.S. military's embrace of high technology. During the 1950s, the services sought new weapons in profusion—jet-propelled fighters and bombers, nuclear-powered submarines and nuclear-powered aircraft, a space plane called Dyna-Soar intended to skip along Earth's outer atmosphere. Not all were built, but they pushed R&D spending ever upward and testified to the commitment of the United States to do almost anything to establish technologically based advantages over the Soviet Union.

The Department of Defense (DOD) and the services steered funds to multiple competing technologies in support of similar or identical missions, whether long-range strikes on Soviet territory or tactical defense of the Navy's ships—a mindset still evident today in the more than 50 antiarmor weapons in the inventories of the four services or under development, even though there is no prospect of confrontation with a tank-heavy army such as the Soviet Union maintained through the 1980s. In the civilian economy, competition stimulates innovation. In defense, competition within and among the military services and between the United States and the Soviet Union drove spending to very high levels. Military innovations resulted from this competition; so did duplication and waste. Civilian officials have tried over the years to rein in and rationalize acquisition. They have had little success.

As the Vietnam War intensified in the 1960s and combat operations consumed more of the defense budget, the era of heroic technical ventures waned. Even so, DOD's commitment to high technology almost regardless of cost remained on display. And during the 1990s, R&D continued more-or-less unchanged even as other components of the defense budget shrank in the aftermath of the Cold War. Military R&D passed the $50 billion level in fiscal 2003 and will exceed $70 billion in 2007. Most of the money still goes, as throughout the past half-century, toward weapons systems that are designed to win high-intensity wars against powerful adversaries with large militaries and sophisticated technical systems of their own, even though no such adversaries are in sight and the nation's latest security challenges have to do with combating terrorism, which calls for good intelligence more than massive firepower, and with small-scale fighting as during the occupation of Iraq.

Some of DOD's early high-technology programs proceeded at a reasonable pace to successful conclusions. Five years after the world's first nuclear-powered submarine, the Nautilus, set its first underwater records in 1955, Polaris submarines carrying ballistic missiles were hiding in the North Atlantic. Other programs, including Dyna-Soar and the nuclear-powered bomber, were canceled after billion-dollar expenditures, for technical reasons or because they could not promise military utility. Some R&D ventures, such as those directed toward stealthy aircraft and precision-guided weapons, continued for many years before reaching fruition. Laboratory research aimed at reducing radar reflections from aircraft began in the 1950s. Full-scale prototypes of stealthy aircraft flew in the late 1970s. Then came the F-117 and the B-2 bomber. During World War II, U.S. forces targeted bridges in Burma with radio-controlled bombs; precision-guided weapons proved themselves in the late stages of the Vietnam War and in the 1991 Persian Gulf War demonstrated the impacts on warfighting for which they have since been so widely acclaimed.

During the Cold War, the armed forces got much of what they wanted in high-technology systems and equipment. If the services did not get everything, that was because their wish lists were almost indefinitely extensible. The Air Force put a relatively low priority on the B-1 bomber and did not fight hard when the Carter administration canceled the program. After Ronald Reagan won the presidency and began pumping new money into defense, the B-1B entered production. The Air Force has already begun retiring B-1Bs, although the planes are not old as military equipment goes. Why? The B-1B is a fundamentally flawed design. As explained in chapter 6, its "architecture" was compromised from the beginning by Air Force insistence on the incompatible requirements of high-altitude supersonic flight and terrain-following bombing runs at treetop level. Because of the technical compromises needed to meet these requirements, almost any mission that might be assigned the B-1B can be better accomplished by some other weapon or platform. Not a few other systems also proved poorly suited to the tasks eventually assigned them, including supersonic fighters in the "Century Series" such as the F-105. During the Vietnam War, the Air Force lost nearly half its inventory of F-105s flying these planes on missions very different from those for which they had originally been designed.

Well before the collapse of the Soviet Union, the B-2 bomber had come to represent a fundamental dilemma of acquisition: the United States was spending more and more money to purchase fewer and fewer weapons systems. The Air Force originally planned to buy 132 B-2s; the number was cut to 75 in 1990 and 21 in 1992, leaving the final cost per plane at $2.1 billion, three-quarters of this representing prorated R&D. With such a price tag, and with peacetime accidents historically claiming more than 50 military aircraft each year, it is hard to see how the Air Force can risk flying B-2s except on critical combat missions.

Even as technology cycles in the civilian economy have shortened, weapons systems take longer to develop, often in excess of a dozen years and sometimes two decades or more. This makes it increasingly difficult to take advantage of commercial innovations or respond in timely fashion to the new or unconventional threats of "postmodern conflict." Design work on the F-22 fighter started in the early 1980s, the first prototype flew in 1990, testing of a production-like model began in 1997, and the first F-22s were declared operational at the end of 2005. Production will extend through 2011 or 2012. In part because the R&D and design stages of such programs drag on much longer than planned, DOD's high-technology systems almost invariably cost more than originally projected, forcing cuts in production to keep total expenditures to tolerable levels. When work on the F-22 began, the Air Force claimed each aircraft would cost about $35 million, including apportioned R&D, enabling it to purchase 762 of the planes. Two decades later, with R&D spending surpassing $20 billion and planned production cut to about 180, each F-22 represents a "program acquisition unit cost" of some $360 million.[2]

Under U.S. law, each service is responsible for its own acquisition programs. In exercising this responsibility, the services maintain tight control over performance requirements and, as in the case of the B-1B, sometimes refuse to trade off requirements against one another or against cost. In other cases, they keep changing the requirements, because the service's "strategic vision" changes or because turnover in the high ranks brings a different personal vision to decisions on weapons purchases. Too often, R&D and engineering continue year after year in pursuit of technical objectives that are unattainable (or barely attainable), unstable, or both. As a consequence, when systems finally reach the field they often include obsolete or obsolescent technology; no matter the effort to keep them up-to-date through "technology insertion" it is impossible to modify everything in a complex system at once. While the weapons system as a whole may have very impressive capabilities, that is only because such enormous sums have been expended. The dynamic is perverse. The push for highly advanced technologies to overmatch any conceivable adversary is largely responsible for acquisition cycles that sometimes stretch beyond two decades. Yet the Pentagon has never found a way around it.

Reforms have been discussed and debated almost continuously since Robert S. McNamara's tenure as Secretary of Defense in the 1960s. Official statements notwithstanding, meaningful changes have been few. That should be little surprise. Large organizations display much inertia and DOD is both very large and split into factions that negotiate deals behind the scenes which they later present to civilian officials and the public at large as fait accomplis. If corporations such as General Motors that have been under intense competitive pressures since the 1970s cannot respond effectively, why should we expect the military, free of such pressures, to react except slowly to the end of the Cold War? Meaningful change in how the nation chooses weapons systems will take a truly determined effort by civilian officials extending over several administrations and congresses.

This book explores the reasons defense acquisition so often seems out of control and resistance to reform has been so persistent and strong. It does so by delving into the institutions involved and how they have changed over time, and into military technological innovation and its necessary accompaniment, doctrinal innovation—the shifts in "military art" that coevolve with technological change. Technology and doctrine are tightly coupled and interdependent. They develop along parallel tracks and must be examined together, just as innovation in the civilian economy must be examined in the context of business strategy and practice.

Most innovations, military or commercial, result from accumulated incremental change. The atomic bomb was the exception, radar, the closest thing to a militarily decisive technology in World War II, more nearly representative. Radar stemmed not from some breakthrough discovery or invention but from ongoing developments in many places by many people beginning in the 1930s, much of it through trial-and-error following fortuitous observations of reflected radio signals in the 1920s. Radar, moreover, emerged as a revolutionary warfighting technology for reasons as much organizational as technical. During the Battle of Britain in 1940, effective procedures for "filtering" information from radar installations along England's channel coast, with the gist transmitted to flight control officers, enabled the Royal Air Force to scramble its fighters in time for pilots to reach attacking positions above incoming German bombers. If radar won the Battle of Britain, it did so as part of a complex system involving the people who manned the Chain Home radar stations and filter centers, planned the organization and its communications channels, flew the planes, repaired them afterward. Too often, high technology is discussed as if its essence could be captured by measures such as Moore's Law, the celebrated relationship showing that the functional performance of integrated circuits (ICs) doubles every 18 months or so. More powerful ICs make possible digital electronic systems of greater complexity. It is the systems, not their components, that are the essence and proper emblem of high technology.

Jet engines, missiles, and computers developed incrementally during World War II and afterward, much as did radar. For computers and their semiconductor components, military and civilian applications were closely related and mutually reinforcing. When the Internet burst into public consciousness in the mid-1990s, it did so following three decades of advances in wide-area computer networks, many of them sponsored by defense agencies and most of them invisible outside the technical communities directly involved. In contrast to the IC chip and the microprocessor, both of which resulted from identifiable acts of invention, the Internet is a radical innovation that coalesced from many incremental innovations. Some of these were predictable, as by Moore's Law; that is to say, they were straightforward extensions of past developments in ICs, computer hardware and software, and telecommunications (e.g., fiber optics, which, like ICs, improved in performance over many years as a result of steady incremental advance). It was their coming together that surprised, and the consequences. The Internet illustrates both the emergent character of many major innovations, including most postwar military innovations, and, a related point that arises repeatedly in later chapters, the extent to which even the most revolutionary-seeming innovations exhibit lengthy prehistories, with precursors that may not be evident at first glance.

Notwithstanding the interdependencies of military and commercial technologies, and the persistence of "spin-off" (from military applications to civilian usage) and "spin-on" (from civilian to military applications), innovation as a *process* differs profoundly between the two spheres. Both technology development narrowly and innovation more broadly are, at bottom, processes of learning. Information and knowledge accumulate, engineers, scientists, managers, and entrepreneurs sort through it, draw tentative conclusions, conceptualize new products and processes, evaluate outcomes, try again. In the civilian economy, market-mediated feedback guides these activities. There is no real counterpart in defense. Wars are intermittent, unpredictable, and unique. The interplay between technology and warfighting practices embodied in doctrine also lacks parallels in the civilian economy. The closer analogies, with the standard operating practices of business firms, are more revealing of difference than similarity. Business routines flourish in stable environments that support repeatable production of standardized goods and services (chapter 8). By contrast, the stock-in-trade of military organizations is the destabilization of operating environments—those of the enemy—through violence, death, and destruction.

When new technologies reach the field militaries must discover how best to employ them, sometimes in the heat of battle. World War I combatants explored what aircraft could and could not accomplish (observation, artillery spotting, ground attack) even as the technology of aviation was advancing rapidly. Much the same was true of tanks, submarines, and poison gas—as it is today for unmanned aerial vehicles, directed-energy weapons, and information warfare. When doctrine for using new weapons is flawed or simply underdeveloped, technologically based capabilities may be moot. The U.S. Army failed to use its Apache helicopters in Kosovo in 1999, in part because it had no settled doctrine to guide operations in concert with Air Force planes but in the absence of the ground forces expected to accompany helicopters into battle to suppress enemy fire and spot targets.

Commercial products, including the long-lived capital goods that most resemble complex military systems (but faintly), face the tests imposed by users and the marketplace every day. Military systems and equipment confront the environments for which they have been designed only in wartime and then only in wars fought against capable and determined adversaries. Feedback-driven learning is a powerful force for continuous improvement in the civilian economy. Feedback into military innovation is sparse and spotty, learning and unlearning discontinuous and error-prone. Simulations, tests, and trials, no matter how painstaking and extensive, cannot reliably predict the performance of a weapon used for the first time in anger against a resourceful opponent with his own ideas of how to fight, his own offensive and defensive doctrine and systems. If and when war does come, moreover, the lessons of actual fighting will often be ambiguous and contested. Should the war end quickly, there will be little chance to evaluate technical modifications or doctrinal shifts until the next outbreak, when circumstances are bound to differ. Given the risks and costs of experiential learning, militaries have good reasons for conservative behavior. At the same time, they have sometimes been prone to unwarranted enthusiasms: for "war-winning" innovations such as precision bombing before World War II; on a more narrowly technical level, for Dyna-Soar after that war; and, much more recently, the National Aerospace Plane, which boosters sold, improbably, to Ronald Reagan's White House as a means of "flying" into Earth orbit.

If military innovation is inherently problematic, the pathologies of DOD acquisition are self-inflicted and therefore avoidable, at least in principle. The government's policies and practices have been shaped by historical experiences rooted in two world wars and the Korean War. During World War I, the U.S. military for the first time confronted, in aviation, high technology in the modern sense. Earlier episodes of military technological innovation, whether on land (mass-produced firearms) or at sea (steam-powered, steel-hulled warships), had been products of cut-and try engineering, much of it conducted by the government's own employees in military arsenals and shipyards. Craft-based methods could not work for the radically new technology of aviation: cut-and-try engineering was a recipe for crash-and-burn disaster. Because neither the Army nor the Navy had internal technical capabilities for design, development, and production of aircraft, the government had no choice but to purchase planes designed and built by private firms. The relationships between Washington and the infant U.S. aircraft industry, and the acquisition policies and practices that grew up during World War I and between the wars, gave birth to what would later be termed the military-industrial complex.

Institutionally, acquisition involves three nested sets of relationships. The first consists of relationships among and within the services—in particular, inter- and intraservice rivalries. For many years, the Army, Navy, Air Force, and Marines have maintained a comparatively stable set of accommodations. (Although administratively part of the Navy, the Marine Corps operates in many respects as an independent service.) They compete aggressively for dollars, the stakes rising when desirable new missions come along or old missions come to seem endangered, as in the second half of the 1940s and into the 1950s, when new technologies and related doctrinal shifts triggered intense conflict over roles and missions.

With the newly created Air Force laying claim to the largest share of acquisition funds in order to build the nation's nuclear deterrent, which at the time consisted entirely of manned bombers, the Navy, having negotiated the transition from battleships to aircraft carriers as the centerpiece of its fleet during the two years or so following Japan's strike on Pearl Harbor, found all of its surface ships, carriers included, seemingly at risk, easy targets for nuclear attack from above or below the ocean's surface provided the enemy could locate them. Although the tests at Bikini Atoll in 1946 allayed some of the concerns, U.S. planners had to presume that the Soviet Union would at some point have enough nuclear warheads to contemplate saturation attacks, while cursory analysis of mutual deterrence suggested greater likelihood of nuclear strikes at sea than on land, where civilian casualties would almost certainly trigger massive retaliation. Because no other nation, and certainly not the Soviet Union, had an ocean-going fleet of much size, there was no opponent for the U.S. Navy to prepare to fight, or to hold hostage. The admirals thus sought to carve out a role for carrier-based planes in a nuclear war. Although they could not match the range or payload of the Air Force's strategic bombers, carrier-launched planes might plausibly threaten the Soviet homeland from the Mediterranean, a body of water sufficient to offer some chance of hiding from the Soviet Union's (land-based) aircraft. The argument was tenuous. It took the Korean War, when the Navy's carriers served as floating air bases to supplement the few and poor airfields available, to demonstrate that naval air power still had a significant military role.

Such paradigm-threatening upheavals have been rare. Since the mid-1970s and the end of the Vietnam War, the service shares of the defense budget have not changed greatly (chapter 3). Despite much talk in recent years of a "Revolution in Military Affairs," based on joint (i.e., multiservice) operations and information and communications technologies for "network-centric warfare," no one seems to contemplate big changes in the roles and missions of the services as a result, perhaps because the services are determined not to allow this.

With reasonably settled interservice relationships, intraservice rivalries, though less visible to outsiders, often have had greater impacts on roles, missions, and force structure. When the United States decided in the mid-1990s to proceed with the F-22 fighter rather than build more B-2 bombers, this was not the result of considered planning. Rather, it was the outcome of an internal struggle over the claims of the two programs on the Air Force share of acquisition dollars.

One of the primary consequences of inter-and intraservice rivalries is to render almost any sort of long-term planning anodyne. As a retired Air Force officer put it, recalling his many years in Pentagon staff positions:

> [L]ong-range planning was an academic exercise only.
>   There were offices with the word plans over their doors and many published documents with this word in their titles, but I was to learn that

none of these had anything to do with the actual decisions made to shape the future Air Force. Instead, these activities usually produced unconstrained wish lists . . . what the Air Force might look like twenty or thirty years down the road if there were no limits on how much money could be spent and no limits on the achievements of science and technology.

The planning world and the budget-decision world were separate . . . .

Budget decisions were based on pragmatic, near-term considerations, such as the degree of political support for a particular program.[3]

As a result, the operative "plan" can only be extracted by looking at how money has actually been spent. That usually reveals, not some more-or-less rational or at least explicable decision process, but the unmanaged outcomes of insider maneuver and political coalition-building.

Unsurprisingly, when the end of the Cold War seemed to call for reassessment of choices among weapons systems, the armed forces could not be impelled to confront the questions raised. Various studies, including Quadrennial Defense Reviews in 1997 and 2001—the latter labeled "pabulum at best"—generated little in the way of meaningful analysis and debate.[4] The most recent review, issued in 2006, places rhetorical emphasis on combating terrorism and irregular forces such as insurgents in Iraq but is nearly empty of specifics (chapter 11). In the absence of broad agreement on future needs, the nation's combat forces have been scaled back among the services and their components in nearly equal proportions: "Since 1990, the Navy has cut the number of battle force ships from 574 ships to 285. However, the composition of the fleet has remained relatively constant, with no one category of ship varying by more than 3 percentage points in its contribution to the total fleet."[5]

The second set of relationships, between military and civilian authorities, also has two dimensions, corresponding to the executive and legislative branches. Within DOD, the services frequently find themselves at odds with civilians in the Office of the Secretary of Defense (OSD). Even so, the armed forces and OSD usually try to present a more-or-less united front to the outside world, including the White House, and frequently succeed. Congress, often accused of meddling, does so in part because it trusts neither the services nor OSD very far. Congressional suspicions of "waste, fraud, and abuse" underlie many of the laws and policies that govern acquisition, and hobble it.

The third and final set of relationships concerns the defense industry and its interactions with Washington. This is the most malleable of the three, although chiefly on the side of industry. While flexibility is a virtue in the civilian economy—if one firm fails after pursuing some innovation that does not materialize, others will take its place—societies expect their militaries to be steadfast and reliable. Military organizations, if too "flexible," might be tempted to adopt new technology or doctrine prematurely, or indeed insert themselves into politics. Defense firms can alter their strategies based on managerial decision; the military is expected to wait on and adjust to political decisions.

The United States could live with conflicting views on what weapons to buy, if uncomfortably, during the Cold War. Complaints of legislative micromanagement notwithstanding, long-running confrontation with the Soviet Union underlay reasonably widespread agreement on national security imperatives. Political compromise often led to purchases of some of almost everything the services wanted. That drove up costs, and on occasion left the nation with too few of the weapons systems that proved most effective, along with others for which it had no real need, but high continuing spending finally helped drive the Kremlin out of the arms race.

The disintegration of the Soviet Union and the crumbling of Russian military power in the 1990s left the United States with no overarching policy or perspective to govern choice of weapons systems. The Pentagon continued to prepare, as the clichés have it, to refight the nation's last war—in this case the Cold War. The services sought and were granted hundreds of billions of dollars for systems that might have been conceived to beat back the Red Army in Europe. Sometimes, they *were* conceived for that purpose, as in the case of the F-22, designed in the early years of the Reagan defense buildup to outpace expected advances in Soviet air-superiority fighters. No one since has been able to articulate a convincing rationale for this very costly airplane.

In 2002, President George W. Bush declared, in *The National Security Strategy of the United States of America*, that "The major institutions of American national security were designed in a different era to meet different requirements. All of them must be transformed."[6] The 2006 rendition of this document repeated much the same message. These and other official statements notwithstanding, changes in weapons system planning that might merit the label "transformation" have been few and modest.

The Soviet Union was a superpower; there has been no superpower other than the United States for the past 15 years, and it is hard to see how such a power could emerge within the next several decades. Some nations have large armies, but none has a military establishment both large and technologically sophisticated. No other nation has a large and capable navy. None can put planes in the air to compare with those of the Air Force or the Navy, much less both (plus the Marine Corps' quite formidable air arm). Any state harboring superpower ambitions would have to make heavy investments over many years. Quite simply, no state that might have the will has the resources, either financial (hundreds of billions of dollars just to make a start) or technological (tens of thousands of engineers, scientists, and technicians with relevant expertise and experience).

While a veneer of "new threats" ranging from cyber-war to terrorism adorned the rhetoric of national security planning in the early 1990s, little hardheaded analysis lay behind it. Sarin gas attacks in Tokyo's subways in 1995 forced governments everywhere to take the new threats more seriously. Nevertheless, they got little money compared to ongoing expenditures for traditional weapons. Following the terrorist attacks on New York and Washington in September 2001, the rise in U.S. defense spending that had begun several years earlier accelerated. Yet terrorism is not purely or even primarily a military threat and provides little basis for making decisions on major weapons programs. While counterterrorism sometimes overlaps with warfighting, the only truly effective defense is to head off terrorists early through preventive action. Given better intelligence, after all, the United States might have forestalled the 2001 attacks and found no reason to send troops into Afghanistan in 2001 and Iraq in 2003.

The 10 chapters that follow seek to explain how the United States arrived at today's weapons acquisition policies and practices, so far as "major programs" are concerned.[7] Without such understanding, efforts at reform will continue to devolve into piecemeal battles over particular weapons systems. That was the pattern during the Cold War, during the dispute between OSD and the Army over cancellation of the Crusader artillery system in 2002, and in 2004 when the Army opted, after an internal struggle (and in the expectation that the funds released could be steered to other Army priorities), to end its Comanche helicopter program, underway since 1983 and finally nearing production.

Chapter 2 expands on the relationships between technology and doctrine. Chapter 3 outlines the penetration of high technology into the military during the 1950s, emphasizing the persistence of the patterns established in the half-dozen years after the Korean War. The fourth chapter locates further precedents as far back as World War I, while chapter 5 describes the post–World War II defense industry and its relationships with Washington.

Chapter 6 includes several brief case studies that illustrate the process of military innovation, one of them dealing with the technological and doctrinal dimensions of air power, a topic that threads through several chapters. Chapter 6 also examines development of the Army's M16 assault rifle, an example showing that, even though most of DOD's more troubled programs involve systems of great complexity, the development of quite straightforward weapons has sometimes been mismanaged too.

Chapters 3–6 give a picture of defense acquisition as it functions today. Chapter 7 then discusses, more generally, the processes of design and development through which ideas and concepts take on concrete, producible form. Chapter 8 extends that discussion to the special problems posed by high levels of complexity—and many military systems exhibit complexities without meaningful parallel in the civilian economy—with particular attention to computer software and the Revolution in Military Affairs.

The next two chapters explore in greater depth how military innovation occurs in practice, through further case studies (chapter 9) and analytical comparison with commercial innovation (chapter 10). The illustrative cases include nuclear power, a commercial innovation that is instructive because misguided government policies short-circuited feedback signals of the sort that ordinarily pace diffusion and deployment in the civilian economy—just the sort of signals inherently lacking in military innovation. The final chapter addresses pathways to reform, arguing that meaningful reform requires an overall shift from military to civilian control and, on the military side of the Pentagon, greater authority for the Joint Chiefs of Staff, and particularly its chairperson. The objective: decision hierarchies in DOD that function more like those in innovative commercial firms.

Aerospace and electronics, including computer software, exemplify military high technology and account for the bulk of acquisition spending. They get much attention in the chapters that follow. Several of those chapters may seem like litanies of troubled programs lightly leavened with a few success stories. The intent is not simply to criticize. Innovation, military or commercial, is unpredictable and ridden with what hindsight reveals to be mistaken judgments. In the civilian economy, failure is inevitable and expected. Corporate managements and financial markets tolerate failures, within limits, scrutinizing them for lessons to be applied in the future. Societies cannot afford to extend much tolerance to failures in military innovation, instead expecting them to be detected and remedied while latent. That is a difficult problem, the more so because the criteria businesses routinely use to evaluate prospective innovations do not apply. It is harder to agree on what constitutes "failure," easier for advocates of this weapons system or that to manipulate bureaucratic and political decisions, and difficult for outsiders, including civilian officials, to pierce veils of obfuscation and secrecy spun by those advocates (in the military or defense firms). A basic contention of this book is that the United States must find better ways of tapping professional military expertise while at the same time ending the monopoly of the services over decisions on major weapons systems. Those decisions should be made in the sunshine, after reasoned and open debate.

Today the United States controls awe-inspiring military power, far beyond that of any other nation. This is a reflection of technological capabilities made possible by an equally awe-inspiring flow of funds. Over the foreseeable future—that is, so long as the United States does not face another superpower or a collection of second-tier powers united in opposition—the nation should have little trouble remaining far ahead of any and all potential adversaries. The United States missed an opportunity for transformation at the end of the Cold War and another opportunity during the presidency of George W. Bush as a result of controversy engendered by the war in Iraq. There will be further opportunities. Policymakers and the public need to be alert for them. Even during the Cold War, U.S. military personnel might have had better systems and equipment if funds had been spent more wisely. The lesson holds doubly today, when there is not even a hypothetical superpower on the horizon, when new threats call for technologies and doctrine quite unlike the major theater wars that continue to serve as the primary justification for DOD's major programs, and when the United States, a Goliath that has left its overmatched adversaries hardly any alternative, finds itself dismayingly vulnerable to unconventional and asymmetric threats.

CHAPTER 2

TECHNOLOGY AND DOCTRINE

Doctrine prescribes and training instills warfighting practices. Doctrine refers to a military's codified prescriptions of how to fight—rooted in experience and adjusted based on lessons from large-scale conflicts such as the Korean War and smaller engagements such as air strikes in the Balkans during the 1990s. In the 1991 Gulf War, doctrine called for the U.S. Army's M1 Abrams tanks to fire on the move, feasible thanks to a computerized fire control system that automatically compensates for motion over rough ground. Because the radar signature of a stealthy F-117 increases by 10 to 100 times in some kinds of turns, pilots train to avoid such maneuvers. These are two simple examples of the ways in which technology and doctrine interact.

This chapter opens the discussion of military technological innovation by sketching the function of doctrine. It also introduces issues such as joint operations and interservice commonality and begins the process of explaining how the United States arrived at today's policies and practices for buying weapons systems.

★   ★   ★

The common perception of the 1991 Gulf War as a victory won by high-technology weapons such as precision-guided munitions diverges from that of analysts and military professionals who submit that doctrine, training, and skill accounted above all else for the ease with which U.S. troops and their Coalition partners drove Iraqi forces from Kuwait. The Defense Science Board contrasted the U.S. experience in the Middle East in 1991 with earlier wars in which poorly prepared troops had to learn to fight while under hostile fire:

> Until 1991 the Army's first battle of each war had been a disaster. In Desert Storm, after a decade of CTC [combat training center] use and with the insistence that every unit that went to war had to do well in the National Training Center, the Army had an overwhelmingly successful first battle.[1]

Military organizations formulate doctrine—warfighting principles and practices, often set down in written manuals—to provide a foundation for the exercise of discretion in combat. Doctrine and discipline instilled through peacetime training serve as antidotes to the confusion and fear of battle, bulwarks against panic and flight. At the same time, doctrine sometimes can become part of a belief system that stifles fresh thinking and inhibits innovation.

Military leaders know that warfighting is chaotic, that participants may have almost no grasp of what is going on around them, that blind terror will sometimes be the dominant mental state among those doing the shooting and being shot at. The fog of war—the disorientation famously portrayed by Stendhal in *The Charterhouse of Parma* and by many others since—obscures events even as they occur. The metaphor extends beyond the smoke that rolled over nineteenth-century battlefields and the difficulty of comprehending what might be happening beyond range of field glass or telescope. Most fundamentally, it refers to the difficulty of making sense of unfolding events in near-real time. No matter the improvements in technical capabilities for remote sensing and surveillance, assembling an overall picture will always require piecing together fragmentary pieces of information (something militaries now call data fusion) from different systems (visual, radar, infrared), weighing ambiguities and resolving contradictions, and, following the battle, extracting lessons from debriefings and after-action reports.

Drill and training, intended from the late sixteenth century to imbue foot soldiers with unthinking obedience, to create automatons prepared to kill and be killed, today are regarded as essential building blocks for small-unit cohesion. Those entering combat need to know not only that they can rely on one another, but also understand more-or-less instinctively how their comrades will react under foreseeable contingencies. Because military effectiveness depends on coordinated action, operations are scripted, planned, rehearsed—and afterward critiqued—literally to the minute, as

suggested in this account of the Marine Corps' amphibious landing on the
Japanese-held island of Tinian in 1944:

> Although the leading wave of troops was dispatched from the line-of-departure
> ten minutes later than planned, favorable conditions of wind and tide enabled
> them to pick up lost time so that it was landed only six minutes behind sched-
> ule. These conditions continued, and the three leading battalions landed
> within two minutes of the prescribed times, and all other waves came shore
> within one minute of their schedule.[2]

Because the plan will sooner or later fall apart (as it did on Tinian), the most
careful preparation cannot substitute for individual volition. Rather, planning, doctrine, and training provide the foundation and starting point for
independent action by people who must cooperate in order to stay alive
and behave in predictable ways to maximize their fighting power.

Doctrine provides general principles that shape planning at strategic
levels as well as tactical. (The term operations is often used to refer to
theater-wide offensives or other large-scale undertakings such as bombing
campaigns that nonetheless constitute only one piece of some larger strategy.) Strategy links military power with political objectives. It is the joint
responsibility of civilian and military leaders, with the final say heavily
weighted toward elected officials and their civilian appointees and military
leaders responsible for ensuring that political leaders understand what the
nation's armed forces can and cannot be expected to accomplish. High-
level strategic considerations enjoined the disposition of the forces assembled by the United States and its allies to drive the Iraqi army out of Kuwait
in 1991. The political goals, at the time, did not extend to the displacement
of Saddam Hussein; in 2003, they did.

During the last two decades of the nineteenth century the naval
doctrine of the United States came to be based on offensive sea power
rather than coastal defense and commerce raiding. Originating within the
Navy—Captain Alfred T. Mahan serving as chief theorist, exponent, and
publicist—sea power gained widespread acceptance by century's end as
properly integral to the strategic posture of an emergent Great Power.[3] The
new naval doctrine revised the template for deciding what kinds of warships the nation should build. After World War II, with no rivals to dispute
control of the seas, and with the Air Force, committed to its own doctrine
of strategic bombing, assuming the role of principal shield against external
threats (through nuclear deterrence), the Navy struggled to redefine its
place. As we saw in the first chapter, it found this in the combination

of power projection ashore from aircraft carriers and later gained its own deterrent in the form of nuclear-powered submarines carrying nuclear-tipped ballistic missiles. With the end of the Cold War, the Navy once more modified its doctrine, which now emphasizes littoral warfare: operations in coastal waters, noisy and confusing environments for sensors (e.g., radar, sonar), dense with civilian vessels ranging from fishing boats and ferries to transiting supertankers, and cramped and confined settings for maneuver. These conditions make different demands on ships and shipboard systems than "blue water" warfare in vast expanses of open ocean, where the first problem is simply to locate the enemy, rather than to distinguish the enemy from noncombatants and background clutter. For some years, the Navy has been arguing that littoral warfare requires new generations of combat vessels, such as the Virginia-class submarine.

Civilian leaders have sometimes accepted service doctrines while hardly realizing their implications. As we will see in later chapters, this was the case for strategic bombing as promulgated by Army aviators before World War II. In effect, both civilian officials and the Army's high command left the Air Corps free to do largely as it pleased. After 1945, when strategic attack came to mean nuclear attack and schoolchildren practiced hiding under their classroom desks, Air Force doctrine became everyone's concern. At the same time, secrecy and tight control by the Strategic Air Command (SAC) over nuclear war plans meant that even the president and his closest civilian advisors did not always grasp the ramifications of decisions they had implicitly or explicitly endorsed.[4]

Developing sound doctrine requires understanding what military systems can and cannot accomplish; the design of those systems likewise requires understanding military preferences and priorities. Design is the core activity of technological practice and system designers must appreciate what the military would like to do, if only it were possible, and also the tasks that political leaders might wish to assign the military, if only the military could accomplish them. Multiple sources of conflict exist, beginning with definition of "requirements"—how the system is to perform—and continuing through the developmental tests that finally determine what the system can actually do, which sometimes falls well short of what it was intended to do.

From the perspective of the services, requirements (or "capabilities"), like doctrine, should be left to them. Military leaders believe that only they can understand what is needed for warfighting, and not, for example, civilians in the Office of the Secretary of Defense (OSD), much less the Office of Management and Budget or the Government Accountability Office (GAO, the congressional oversight agency previously known as the

General Accounting Office) or nongovernmental watchdogs.[5] This is a generally valid claim too frequently wheeled out to deflect scrutiny. It is true that relatively small differences in the performance of military systems may have life-or-death consequences, a compelling concern that too easily lends itself to the rationalization of arbitrary and unrealistic requirements. It is also true that the services may not agree internally on what is needed for warfighting. These disagreements, often pitting staff and field officers against supply and support organizations such as the Army's Materiel Command, rarely have much visibility to outsiders.[6] In the past, when few military officers had technical training or experience, the combat branches sometimes pressed for exotic, high-performing weapons based on unproven technologies while the supply and service bureaus tended to be stodgily conservative and resistant to innovation, particularly from outside. With the U.S. military's postwar embrace of high technology, technical competence improved. Technological optimism has displaced much or most of the opposition to innovation earlier displayed, as illustrated by the lengthy history of R&D on ballistic missile defense.

Requirements set down how the weapon or other system is to function, as specified, for example, by an airplane's speed, range, and armament. Conceptual or preliminary design choices based on these requirements, the choices that define the overall configuration of the system—its architecture—largely determine the system's functional performance and life cycle costs. Design decisions are sequential, tightly coupled, and path dependent. Detail design follows from conceptual design, and once the preliminary choices have been made reversing them carries heavy penalties in terms of budget and schedule. Good early design decisions made the B-52 a good airplane; conflicts among the requirements imposed by the Air Force on the B-1 led to a poor airplane. When Boeing designs a commercial plane, it avoids or adjusts to such conflicts. The critical decisions take place within the firm and can be revisited as necessary during preliminary design. When it undertakes a military program, responsibility for the critical decisions is shared with government and the appropriate set of tradeoffs become a matter for negotiation.

Technological uncertainty and its influence on design decisions should not be underestimated. There could be no guarantees that a stealthy airplane would actually be able to hide from radar until Lockheed built two full-scale demonstrators under the Have Blue program, which started in 1976, some 25 years after the beginnings of R&D on stealth (chapter 7). Have Blue flight tests provided technical knowledge that fed into the design of the F-117. The F-117 did not conclusively demonstrate its effectiveness until successfully penetrating Iraqi air defenses during the 1991

Gulf War. In other cases, uncertainty ends in belated recognition that a long-awaited new technology does not, in fact, perform as hoped, by which time many billions of dollars may have been spent. Uncertainty, finally, means that ineffectual systems can sometimes deter because an adversary cannot be confident that a weapon will not work as advertised.[7]

By definition, innovations push into partially explored territory, meaning that uncertainty is unavoidable. A second difficulty, more serious for military than for commercial undertakings, is that of judging outcomes. No one can say how the attributes of weapons should be measured and compared, what "performance" means under battlefield conditions. Assessments depend on multiple but incommensurable parameters—for example, protective armor versus agility for a tank. In commercial innovation, technical measures may be equally incommensurable. How significant is fuel economy compared with a quiet ride to car buyers? The answers eventually come from the marketplace. In the short term, they may be unclear; nonetheless, they provide a basis for further decisions. There is no analogous basis for designing military systems. Experience helps untangle the meanings of performance, but the lessons of combat may not be conclusive and are rarely timely.

Those lessons, moreover, reflect partial, noisy, and otherwise imperfect records that have little in common with the data-rich settings of firing ranges and field exercises. Thus videotapes shot by civilian bystanders helped resolve controversy over the performance of the Patriot antimissile system during the Gulf War.[8] This example illustrates not only the difficulty of assessing the actual capabilities of military systems but uncertainties that sometimes persist long after they have reached the field: Patriot had been in the Army's inventory for half a dozen years when it faced and failed its initial trial by fire.

Given an ambiguous past and unknowable future, predictions sometimes prove horribly wrong. Hardly anyone anticipated the bloody impasse of World War I trench warfare. Tanks were conceived to roll over the barbed wire and machine guns of stalemated battlefields on the Western Front. The first examples, built by the British, were slow, broke down frequently, got stuck. No one knew how to employ them to best advantage. In his classic study *Ideas and Weapons*, I.B. Holley argued that the U.S. Army was halting and indecisive in exploiting air power during the same war because it had no accepted doctrine to guide use of aircraft, hence to set directions for design and development, which proceeded at a rapid pace as the war continued.[9] By contrast, once the British had tanks in numbers, learning on the battlefield proceeded relatively quickly.[10] Workable tactical recipes were a direct result of the arch-imperative of ending the deadlock. There was no such imperative in the air. Although tanks did not prove decisive during World War I, the British learned to use them within their technological limitations.

As the capabilities of armored vehicles improved over the interwar years, uncertainty again built, and even more as concerned air power. With another war widely assumed to be coming, strategists debated whether to integrate infantry and armor and if so in what proportions, navies built aircraft carriers and experimented with them in fleet exercises, planners asked how many bombers and pursuit planes to buy. Spain's civil war answered only a few of the questions; the rest had to wait. Then, five years after the end of World War II, the U.S. military had to relearn some of the lessons of that war in Korea. Once again not all stuck and had to be rediscovered in Vietnam. The Air Force, notably, lost sight of the need for fighter planes suited to aerial combat at close quarters in its enthusiasm for the technological breakthroughs of supersonic flight and air-to-air missiles (chapter 6). And even as the United States spent billions of dollars on electronic countermeasures to jam and spoof North Vietnam's antiaircraft radars—American pilots sometimes complained that their planes carried so many electronic warfare pods that there was no place left for bombs—more U.S. planes were being damaged or destroyed in ground attacks on poorly protected air bases than were shot down by Soviet-supplied surface-to-air missiles.[11]

Peacetime trials cannot guarantee against worst-case outcomes in which an adversary learns what U.S. weapons can and cannot do and finds ways to defeat them. No one could be sure that the modifications to the Patriot in the aftermath of the Gulf War had substantially enhanced its ability to intercept enemy missiles (missile technology had also advanced), as opposed to test articles on the Army's firing ranges. During the 2003 invasion of Iraq, ostensibly improved Patriots downed two friendly planes.

Since Vietnam, the U.S. arsenal of weaponry has not faced the tests posed by a skilled and committed adversary over an extended period. With exceptions such as the air defenses around Baghdad, the Iraqi military proved too weak to put much stress on U.S. systems (and weaker still in 2003); the tests failed by Patriot in 1991 were easy ones. And because both Gulf wars were short, like other engagements of U.S. forces since Vietnam, the enemy had little time to experiment with ways of defeating U.S. systems. An absence of warfighting experience is something to be thankful for. Nonetheless, it leaves uncertainty concerning both the performance of weapons systems and the appropriateness of doctrine. And even when the experience base is ample, the fog and friction of war and the self-interest of those with reputations and careers to protect mean that retrospective analyses may fail to identify the contributions of technology. Innovations perceived as threatening service roles, missions, or self-image are especially hard to weigh.[12]

Actors and orchestras rehearse and then perform. Military organizations are unique in practicing constantly for conflicts that may never arise. In both military and civilian organizations, learning goes on continually through instruction, training, and repetitive practice. This is true for hospital technicians and insurance agents, who deal with steady streams of new or revised products, and for fighter pilots and intelligence analysts. In all these cases, much or most learning is experiential, taking place "on the job." We expect a surgeon to exhibit greater proficiency after 1,000 heart operations than after the first 100, and a pilot with 3,000 hours to be better than one with 300. The difference is that learning in the civilian economy is "real" in a sense that peacetime military drill and training cannot be. On-the-job learning in civilian occupations is almost entirely informal and undocumented, taking place during the ordinary course of work. Whereas firms sometimes give "raw recruits" in the form of newly hired employees meaningful tasks after a half-day of training, military personnel may retire without ever seeing combat.

When the leaders of the Army Air Corps formulated air power doctrine in the 1920s and 1930s, they made strategic bombing the top priority. (The Army Air Service, after several earlier name changes, became the Air Corps in 1926, the Army Air Forces in 1941, and gained in 1947 the independence aviators had been seeking since the close of the World War I.) In their eyes bombing of strategic targets—e.g., industrial plant and equipment far behind the fighting front—had precedence over all other missions. By degrading the enemy's ability to carry on the fight and the morale of the civilian populace a war might be won through "surgical strikes" with no need for the grinding confrontations between mass armies of World War I. For accuracy, the bombing would have to be carried out during daylight hours in clear weather. (Accurate bombing at night or through cloud cover did not become possible until the advent of radar bombsights capable of useful discrimination among ground features.) The bombers would have to carry machine guns and armor for protection against enemy interceptors and fly at high altitudes to reduce the dangers of antiaircraft fire. All these stipulations had their rationales, but none was based on empirical evidence. The Air Corps could not know whether their preferred doctrine would achieve strategic objectives, or indeed militarily useful objectives of any sort. There had been no meaningful tests during World War I, when aviation technology in any case had been primitive (and the results achieved during World War II would in fact generate controversy for years afterward).[13]

The Army Air Corps was not alone in its belief in strategic bombing as the cardinal application of air power, indeed as a war-winning weapon on its own. But more than any other nation, the United States channeled its available resources to long-range heavy bombers such as the B-17 and B-24 at the expense of other aircraft. The first point here is simply the difficulty in learning faced by military organizations in peacetime. As we will see in later chapters, there is a second point: when evidence began to come in during World War II indicating the limitations of their doctrine, the Army Air Forces first disregarded it and then adjusted only grudgingly under the pressure of unsustainable loses of men and planes. All the while, from the 1920s onward, they neglected tactical aviation.

Military organizations discuss and debate doctrine, formalize it in writing (and today in the computer code integral to many weapons systems), inculcate it through training and maneuvers. Each of the services does this in its own way. Before lightweight and reliable two-way radios, armies had little choice but to delegate extensive authority to field officers and decentralized command and control remain the rule; only if robots replace soldiers will commanders be able to move troops as if they were chess pieces. By contrast, shipboard communications have always been robust (although not communications between ships), enabling the commander to exercise close control over everything that took place on board. And because a ship at sea is in constant "real world" operation, navies traditionally relied on shipboard training, with less use of handbooks and manuals than did armies. As a result, operating practices such as those for carrier take-offs and landings have tended to vary somewhat from ship to ship.[14]

Joint or combined operations, the hallmark of modern militaries, depend on coordination among combat branches (e.g., tanks and infantry) and services (e.g., air support of ground forces). Over the lengthy history of military organizations, joint operations are a recent innovation. During World War I, weapons capable of sweeping mass armies from the battlefield with appalling casualty rates called up responses that demanded close coordination between infantry and artillery so that small units could advance under covering fire (tanks came later).[15] World War II opened with German blitzkriegs integrating attack planes and armor with infantry. Amphibious landings mounted by the Marine Corps and Navy in the Pacific became the notable U.S. innovation in coordinated command and control. With the Army and Navy unwilling to work together, the Marines had claimed the amphibious landing mission in the 1930s. If the Army could not collaborate on doctrine and tactics with the Navy, the Marine Corps, attached to the Navy since 1798 and on perennial alert against absorption into the nation's other ground forces (or reversion to guard and garrison duties), had no such reluctance (the Navy was less enthusiastic). In doing so, the Corps managed to establish itself as a separate if not quite coequal service.[16]

During the interwar years, advances in aviation had begun to erase the centuries-old boundary between war on land and war at sea by giving land-based forces the ability to hunt ships from the sky and navies the ability to launch air strikes on targets many miles inland. Even so, the U.S. Army and Navy mostly fought their own wars in the Pacific, as they did a quarter-century later in Vietnam. During the 1983 U.S. invasion of Grenada, "incompatible radios prevented Army troops from calling in air strikes by Marine aircraft"; more seriously, "a weak [Joint Chiefs of Staff] was incapable of decisive organizational and strategic action."[17] The most significant reorganization of the Pentagon in recent decades, a product of the 1986 Goldwater-Nichols Act, responded in part to these indications of disarray. Improvements were modest. In the 1991 Gulf War, "the Army and Marine Corps efforts represented, to all intents and purposes, independent campaigns."[18] The services evidently attained a somewhat higher degree of integration during the 2003 invasion of Iraq.

Even the purely technical issues have been difficult. In the early amphibious assaults of World War II, Marine riflemen, lacking radio frequencies shared with tanks, signaled by pounding on tank hulls with spent shell casings. Half a century later during the 1991 Gulf War, the Air Force and Navy had no common secure communications channels, so that hundreds of pages of Air Tasking Orders, specifying everything from targets and timing for each sortie (a sortie is one flight by one plane) to the nearly 1,000 radio frequencies assigned to Coalition aircraft, had to be flown as hard copy each day from Riyadh to the Navy's carriers.[19] Afterward, the need to resolve such problems became conventional wisdom. Even so, GAO reported a decade later that "in a battle situation, the Joint Forces Commander is faced with integrating, in an ad hoc manner, more than 400 different mission and software applications."[20] And in Iraq in 2003, "interoperability problems continue[d] to plague tactical communications and contribute to friendly fire casualties."[21]

Effective integration also requires a coherent body of shared doctrine, cemented through joint training, and an unambiguous chain of command (the need for "unity" in command and control is a staple of military thinking). These matters are far less straightforward than technical issues such as radio protocols, encryption standards, and compatible aerial refueling rigs. Each service, and each service branch, has its own culture, a matter of shared beliefs rooted in history, tradition, and self-image. These are sources of organizational strength, foundations for doctrine, and also sources of parochialism.

Given that innovations, by definition, disrupt existing organizational routines, forcing changes in behavior and often in mindset, the social learning that accompanies innovation often proves more disruptive, time-consuming, and costly than the technological learning (for business firms as well as military organizations).

There is perhaps no better illustration of the interplay among technology and doctrine than air support of ground forces, close air support especially—attacks on enemy forces that may be within a hundred yards of friendly troops. While effective air support requires constant communication and careful coordination between air and ground, the mission has been bedeviled by fractured command chains from the beginning. Army aviators have had little use for the mission—when the leaders of the Army Air Corps embraced strategic bombing between the wars, they went to far as to declare ground support a misuse of air power (chapter 6). At the same time, they were unwilling to give it up; if and when such missions were conducted, aviation officers had to control them. The postwar Air Force remained reluctant to purchase planes and train pilots to support the Army's ground troops, yet would not abide the Army taking back the mission. The 1948 Key West accords negotiated among the services barred the Army from operating any except unarmed fixed-wing aircraft. But the agreement said nothing about helicopters. Over the next decade or so, gas turbine engines, much lighter than piston engines of the same power, made it possible for helicopters to lift heavier loads. They could now be armed and armored. Every organization that hopes to innovate confronts the issue of encouraging creative ventures at lower levels while imposing some sort of unifying perspective from above. In this case, technological change in the form of gas turbine-powered helicopters inspired a group of mid-ranking Army officers to initiate a quiet collaboration with a loose collection of defense firms, resulting in the "invention" of the helicopter gunship.[22] With its new helicopters, the Army recaptured a portion of the close air support mission. During the Korean War, helicopters served mostly as air ambulances and utility vehicles for battlefield transport; in Vietnam, they became mounts for "air cavalry," a new type of weapons platform under the control of the Army's own command hierarchy.

By many accounts, the resulting combination of helicopters and Air Force (and Navy and Marine Corps) fixed-wing attack planes represents a far from ideal mix. Skeptics have suggested that helicopters, given their vulnerability to ground fire, are fundamentally second-best weapons platforms. (The Army lost five times as many helicopters in Vietnam as the Air Force did fixed-wing planes; in Iraq in 2003, Marine Corps helicopters operating under doctrine that allowed them to maneuver while firing their weapons took fewer hits from the ground and suffered much less damage than Army helicopters, which were expected to hover before firing.) Advocates had not tried to establish a stable role for helicopters in Army doctrine, in part because the officers originally responsible, who cultivated backing where they could while avoiding unwanted attention, neither sought nor received endorsements from the top levels of the service. Lack of fully formed doctrine was one reason why Apache helicopters did not participate in attacks on Serb forces in Kosovo in 1999. Having shipped 24 Apaches from Germany (following a presidential decision), along with 32 support helicopters, 5,000 Army personnel, and 15 M1 Abrams tanks, plus assorted antitank weapons, howitzers, and rocket launchers (a level of support and protection that required over 500 C-17 cargo flights and seemed to some observers to be foot-dragging by those opposed to the deployment), the U.S. military quite literally did not know what to do with them; time ran out before decisions were reached.[23]

Until the 1958 Defense Reorganization Act created the post of Director of Defense Research and Engineering (DDR&E) within OSD, the services exercised nearly complete control over both R&D and procurement. By placing the DDR&E on the same level as the service secretaries, Congress gave civilian officials substantial leverage for the first time over weapons design and development. When Robert McNamara became Secretary of Defense in 1961, he used this leverage to exert what has since become near-constant pressure for the services to reduce duplication and overlap in their programs and agree on commonality and interoperability. McNamara and his successors have had only limited success. In 1994, for example, OSD killed the Tri-Service Standoff Attack Missile because it seemed impossible to meet all the requirements imposed by the three participating services. These called for a missile that could be carried by any of eight different aircraft and able to accommodate any of five different warheads, objectives that were nowhere in sight after a decade of work and $4 billion in expenditures. Meanwhile, projected unit costs had risen above $2 million per missile, which would have precluded use against any but high-value targets. The follow-on program, restricted to the Air Force and Navy, soon began to miss its budget and schedule targets too.[24] In an earlier and better-known example, McNamara had compelled the Navy and Air Force to share the TFX/F-111 fighter-bomber with minimal modifications for their respective operating environments and missions. Those proved too dissimilar: the F-111 did not serve either very well.[25] On the other hand, the F-4, a Navy plane reluctantly adopted by the Air Force during the Vietnam War, performed well for both services (and for the Marine Corps) over many years. In this case, heavy losses of Air Force planes such as the F-105 in the early years of the war combined with OSD pressure to compel the Air Force to buy planes originally designed for the Navy. In a more recent effort at commonality, the Air Force, Navy, and Marines will each get their own versions of the F-35 Joint Strike Fighter. The Marine Corps, light on tanks and artillery compared to the Army and thus more dependent on airborne firepower, wants planes that can take off and land nearly vertically from unprepared areas close to the front lines. The Navy needs structural strengthening to withstand carrier takeoffs and landings. As a result, considerable cost and complication have been added to what had originally been intended as a relatively simple and inexpensive airplane.

The Revolution in Military Affairs (RMA), which promises to make a great deal of battlefield information continuously available to small units and even to individuals, calls for unprecedented levels of interservice cooperation. Since the 1960s, aerial and satellite surveillance have removed many of the potential hiding places for enemy forces, during daylight hours especially, except for those beneath the surface of the earth or the sea (and even these can be penetrated to some extent). The next hurdle is rapid and effective *use* of this information for command and control in near-real time. Delays of hours, sometimes days, still mark the processing of much intelligence information and surveillance data, meaning that long before targets for precision weapons have been verified they may have moved. The technical obstacles will probably be overcome before those of doctrine (chapter 8). Robotic aircraft operated by the Air Force, for example, might transmit images simultaneously to Army helicopters and ground units and to loitering Navy strike aircraft. Sighting an enemy tank, is an Air Force unmanned combat air vehicle to launch one of its own missiles or defer to one of the "assets" of some other service? A host of such contingencies will have to be harmonized through jointly developed doctrine.

Beyond such matters, what kind of equipment might U.S. forces need for further fighting in the Middle East? For peacekeeping missions and counterterrorism? How might the RMA, originally envisioned as a means of "managing" major theater war, contribute to the small-scale conflicts in which U.S. forces are likely to find themselves embroiled in the future? There are no clear or agreed answers to such questions. Yet the United States seems quite prepared to spend ever greater sums on new weapons without much attempt to find answers.

CHAPTER 3

KOREA AND ITS AFTERMATH: THE SHIFT TO HIGH TECHNOLOGY

World War II tanks retained as monuments on pedestals at Fort Knox had to be refurbished and issued for combat [in Korea].

—Philip L. Bolté, "Post–World War II and Korea:
Paying for Unpreparedness"[1]

More than World War II, the Soviet Union's nuclear bomb tests, or Sputnik, the Korean War, in which outnumbered U.S. forces were nearly pushed into the sea, transformed the U.S. military's approach to R&D and procurement of weapons. With the surrender of Japan, the Truman administration slashed defense spending. Korea reopened the budgetary taps and initiated a period of heroic technological ventures. A continuing push for superior weapons to offset the numerically larger ground forces of the Soviet Union replaced the attritional strategies of past U.S. wars.

World War II began the transformation of military attitudes toward technology. The armed forces had earlier viewed research much as they did industrial mobilization. In an emergency, factories would convert, making boots and tanks instead of shoes and cars; engineers and scientists would turn to war-related technical work. Beginning in 1940 and accelerating after Pearl Harbor, temporary structures for military technology development and production were thrown up, as they had been a quarter-century earlier during the World War I, to supplement the military's own arsenals, supply bureaus, and shipyards. These structures melted away following the surrender of Germany and Japan. There was little money for new weapons until the Korean War, when outnumbered U.S. troops fighting with obsolete equipment suffered heavy losses. Near-disaster in Korea deeply affected policymakers in Washington. Military R&D rose steeply over the second half of the 1950s and has remained high ever since. By the end of that decade, the United States had put in place a new set of institutions to support the design, development, and production of high-technology weapons. These structures have changed relatively little since the late 1950s.

This chapter, and the several that follow, describes both prewar and postwar institutions, while making no attempt to do so exhaustively: the objective is to provide context for understanding military technological innovation and explain how the United States ended up with today's acquisition policies and practices.

The Lesson of Korea

The Cold War arms race had two major precedents: the buildup of naval power before World War I and rapid developments in aviation technology during the 1930s. The naval arms race that began in the latter part of the nineteenth century was in considerable part a race to spend money, the more so once the pace of technological change slackened and competition took the form of a race to build ever larger, ever more costly all-big-gun battleships. Often called, eponymously, dreadnaughts, after the first of its kind, launched in Britain in 1906, these were technically sophisticated for the time, notably in their centralized fire control systems, but otherwise basically exercises in engineering bigness. Because it was possible to build fast, heavily armed and armored battleships, they had to be built. Doctrinal debates centered on how big to make the guns (the biggest guns could not be fired as rapidly as those of somewhat smaller caliber), how thick the armor, how powerful the engines more than on how these vessels would fight.[2] The chief mission of the battleship, inheritor of a tradition going back to the heavily gunned ships-of-the-line at the core of naval power since the seventeenth century, had always been to fight the enemy's biggest ships. That mission remained paramount through World War I, but was superseded in World War II when aircraft carriers and naval aviation performed it better.

The interwar arms race in aviation had both technological and doctrinal elements. It exhibited a wider range of national choices and less of the emulation that characterized the pre–World War I naval buildup. Aviation technology was still quite primitive at the time of the 1918 Armistice. Rapid technological advances during the 1920s and especially the 1930s would in any case have rendered lessons drawn from the 1914–1918 war

largely moot. With nothing better available, "theories" of air power
exerted heavy influence over the choices made by major powers, not least
the United States. As we saw in the preceding chapter, according to Army
Air Corps doctrine strategic bombing would destroy the enemy's ability to
fight by crippling his industries and his will to fight by demoralizing the
civilian populace. As later chapters relate, World War II proved the faith of
Army aviators to be ill-founded. The reasons included respective rates of
technological change for bombers and pursuit planes. Heavy bombers were
ahead for a few years in the early 1930s in the United States, in part because
of the stimulus to innovation provided by airline demand for planes suited
to long-distance flights between major population centers. In the second
half of the 1930s, pursuit aircraft overtook bombers in performance.
Constrained to fly a straight and level path if they were to have any hope
of delivering warheads somewhere near their targets, heavy bombers were
now vulnerable to both pursuit planes and ground-based antiaircraft fire.
Five decades would pass before stealth enabled bombers to hide from
enemy radars, interceptors, and antiaircraft missiles.

Although World War II was basically one of attrition, much money
went for R&D (figure 1). Radical innovations were uncommon, the atomic
bomb the principal exception. Radar too counts as a radical innovation, but

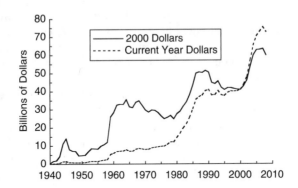

Figure 1   U.S. Military R&D

Note: 1940–1948 includes the National Advisory Committee for Aeronautics, which
worked almost exclusively on military projects during this period; later years exclude the
National Aeronautics and Space Administration, for which no military-civil breakdown is
available. 2007 and 2008 estimated.

Source: 1940–1948—*Federal Funds for Science XI: Fiscal Years 1961, 1962, and 1963*, NSF
63–11 (Washington, DC: National Science Foundation, 1963), Table C-32, p. 136;
1949–2008—*Historical Tables, Budget of the United States Government: Fiscal Year 2008*
(Washington, DC: U.S. Government Printing Office, February 2007), Table 9.7.

it was one that emerged gradually rather than suddenly, the result of a continuing stream of mostly incremental advances (with microwave radar a major exception). Radar altered warfighting practices in many ways. Ships at sea could no longer hope to hide in the nearly empty ocean, as they had for centuries. Airborne radars sensitive enough to detect submarine snorkels became available, a vital boost in protection for Allied shipping in the North Atlantic. In the Pacific, proximity fuzes based on miniature radars built into the nose of antiaircraft shells made the U.S. Navy's guns, by then radar-aimed, far more deadly. Because these fuzes detonated the explosive charge upon sensing a nearby object, gun crews no longer had to hit a plane to bring it down, a nearly impossible feat until the attacker was on top of them. "By 1945 an airplane caught in the range gate of an SCR-584 radar feeding the data to an M-9 director controlling an automatic tracking 90 mm gun using proximity fuzes simply meant that the plane was finished."[3]

Other wartime innovations came to full fruition well afterward. German rocketry laid foundations for the missile age, but the V-1 flying bomb and ballistic V-2 had little impact on World War II. Jet fighters reached the Luftwaffe in 1945. They could climb faster to higher altitudes and reach greater speeds than anything the Allies could put against them (and had no need of the high-octane gasoline that strategic bombing had begun to deny the German air force). The new jets caused much concern among Allied air commanders, but Germany did have enough of them to make a difference. The Allies won by outproducing the Axis, rather than seeking war-winning innovations, the German strategy, which was poorly executed and failed.

During the first half of the war especially, the United States rarely had the best weapons. German tanks were better, at the end of the war as well as the beginning. "The Japanese submarine torpedo was far superior . . . ." while "German submarines had better surface and submerged speed and superior sonar, optics, diesel engines, and batteries. They could dive deeper and faster."[4] But labor productivity in U.S. defense industries exceeded that in Germany by a factor of two, that in Japan by a factor of five.[5] In December 1942, U.S. shipyards launched more cargo vessels and tankers than in all of 1941. From 1942 to 1945, the U.S. Navy commissioned 28 large (fleet) aircraft carriers; Japan could launch but 10. By itself, the United States produced more small arms and artillery pieces than the Axis nations in total.[6] The Manhattan Project too depended on a production effort of mammoth proportions to yield the few hundred pounds of fissionable material, uranium-235 or plutonium, needed for a handful of bombs.

Military R&D was nearly invisible in the United States before the war and then fell back to low levels along with the rest of the defense budget, as figure 1 shows, even though the Pentagon by the end of 1945 had begun making contingency plans for war with America's erstwhile ally, Soviet Russia. The big jump in R&D came after the Korean War. During the 1950s, RDT&E (research, development, test, and evaluation, the Pentagon's label) grew by seven times in nominal dollars and five-and-half times in real terms. Most of the increase came in the second half of the decade. During World War II, the U.S. military had begun to define a new set of relationships among R&D, weapons design, and doctrine. During the second half of the 1950s, these relationships were further developed, refined, and consolidated.

The unexpected plunge into a major new conflict in Korea, and especially the retreat of U.S. troops in late 1950 following the entry of Chinese forces, deeply alarmed official Washington. The nation's military had contracted sharply after World War II. Battling over roles and missions keyed to the prospect of all-out nuclear war with the Soviet Union, the services had not been thinking about, planning for, or preparing to fight a limited war on the ground, tightly constrained geographically and with cloudy geopolitical objectives. The United States would be similarly unready, in terms of its military thinking and to considerable extent in weaponry, for the war that came along in Vietnam a dozen years later.

The lessons U.S. policymakers drew from Korea began with the sheer weight of opposing forces. If China sent troops across the Yalu in the autumn of 1950 unprepared for winter weather, with no tanks and little artillery, the USSR would be a formidable opponent and many in Washington worried that North Korea's initial invasion had been a feint, a prelude to Soviet incursions elsewhere. The Red Army knew how to fight, the Soviet Union had long since proven its technological capabilities, and Kremlin leaders had demonstrated their indifference to casualties in World War II, when perhaps 10 million Soviet troops lost their lives (and by many accounts an even greater number of civilians). America's Cold War planners, deeply concerned since the close of World War II with the ground strength of the Red Army and its presumed ability to sweep through Western Europe should some miscalculation trigger hostilities, had no wish to contemplate war against such an opponent without overwhelming qualitative advantages.

Creating those advantages would not be easy. The Joint Chiefs of Staff had been persuaded even before the 1948 Berlin crisis that "imbalance between forces and commitments" meant "a substantial degree of rearmament was imperative."[7] After Korea, it seemed plain that greatly increased

R&D would have to accompany rearmament to provide U.S. forces with the best possible weapons. Stalin's generals had fought World War II not only by throwing men at machines but by throwing machines at machines. Russia's factories had turned out more airplanes than Germany's (137,000 compared to about 110,000).[8] Russian tanks had proven a match for those of the German army. The Soviet military had not fully demobilized after the war and in the late 1940s began to expand and modernize.[9] Technological advantages for the United States would have to extend well beyond an ever-expanding stockpile of atomic bombs, given that the Soviet Union had demonstrated its own fission weapon in 1949, and in 1953 tested a small fusion device, much earlier than expected by U.S. intelligence and only nine months after the first U.S. test of a hydrogen warhead.[10] Soviet design bureaus had long since copied the B-29: Washington policymakers knew in the early 1950s that the USSR had bombers (though not how many) able to reach the United States, albeit on one-way missions.

It had been plain long before the 1957 Sputnik flights that nuclear warheads could be combined with unmanned aircraft modeled on Germany's V-1s, or ballistic missiles, descendants of the V-2. Immediately after the war, the U.S. Army brought V-2 technical director Wernher von Braun and a number of his coworkers to the United States, along with a large store of missile components, and set them to work. In 1946, Vice Admiral Arthur Radford stated the intent of the United States "to take full advantage of scientific research and development . . . including guided missiles and pilotless aircraft."[11] Engineering development of cruise missiles began shortly thereafter. While some Americans, including Vannevar Bush, head of the World War II Office of Scientific Research and Development (OSRD), argued that missiles could not be made accurate over multi-thousand mile ranges, nuclear warheads, fusion bombs especially—a thousand times more destructive than the warheads that devastated Hiroshima and Nagasaki—rendered high levels of accuracy moot, at least as far as attacks on cities were concerned. What the United States could do, the Soviets presumably could do too. The continental United States, earlier insulated, even isolated, by oceans, the arctic, and pacific neighbors, would in the future be open to devastating attack from the sky. Then in 1957 the first two Sputnik launches showed the entire world that the Soviets had, or soon would have, the ability to target the United States with intercontinental ballistic missiles (ICBMs).

The USSR's nuclear weapons tests had been expected; only the timing surprised. The public reaction to Sputnik was one of shock verging on panic. Still, so far as U.S. policy goes, this was the second shoe to drop. The Korean War had already sparked decisive change in national security policy.

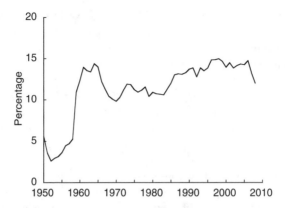

Figure 2   Defense R&D as a Percentage of U.S. Defense Spending

Note: 2007 and 2008 estimated.

Source: Historical Tables, Budget of the United States Government: Fiscal Year 2008 (Washington, DC: U.S. Government Printing Office, February 2007), Tables 3.1 and 9.7.

Time magazine called the rout of U.S. forces at the end of 1950, capped by the decimation of the Eighth Army in December, "the worst defeat the U.S. had ever suffered."[12]As the lessons of Korea were absorbed, extended, and sometimes embroidered, the U.S. push into high technology military systems broadened and deepened. Defense spending declined following the 1953 armistice. The R&D component in defense spending did not. Instead, it rose.[13] As the U.S. military sought in the high technologies of the day what would later be termed "force multipliers" to enable it "to fight outnumbered and win," RDT&E accounted, by 1960, for over 12 percent of the defense budget, compared to about 3½ percent in 1955. As figure 2 shows, in only one year since, 1970, with funds drained from RDT&E to pay for the war in Vietnam, did the share slip below 10 percent.

Technology and Procurement before and after World War II

Until World War II, most U.S. government research funds went for agriculture. Otherwise, public policies had not created an institutional framework and experience base for innovation extending much beyond the patent system, enshrined in the Constitution, direct grants to inventors made from time to time by Congress, and support for land-grant colleges under the 1862 Morrill Act, a policy itself closely linked with promotion of agriculture. During the early part of the twentieth century, industry moved far ahead of government in learning to manage and exploit technology.

Industrial research, identifiable as such, emerged in the last two decades of the nineteenth century in Germany. In the United States, companies such as General Electric, Du Pont, and AT&T began to make a place for research in corporate strategy. During World War I, synthesis of poison gases and explosives demonstrated the utility of science-based research in ways that profit-seeking firms could hardly miss.[14] By the early 1920s, some 500 American companies reported that they operated research laboratories. While most of these facilities were small, with fewer than 50 employees, both numbers and staffing rose steadily through the 1930s, despite the depression.[15] In industries such as automobiles, American manufacturing firms looked to R&D to differentiate their products, to reduce production costs (e.g., by finding ways to stamp fenders with fewer dies), and for innovations with customer appeal (automatic transmissions, high-compression engines for greater power). Meanwhile the National Bureau of Standards (NBS, since 1988 the National Institute of Standards and Technology), established by Congress in 1901, remained almost alone as a (nonagricultural) government research organization.

At first, industrial laboratories worked mostly on production problems, troubleshooting factory processes, testing purchased inputs such as paints and electrical insulation, monitoring quality. Over time, research contributed a growing number of product and process innovations, many of them the results of incremental improvements to outside inventions purchased by the firm.[16] The U.S. government made no effort to follow a similar path. The military had little technical capability except in armaments and warships (and the infrastructure work assigned the Army Corps of Engineers, such as water projects). While the potential of the airplane as a weapon of war had been evident when the Wright brothers first flew, the Army and Navy hesitated to buy planes.[17] Since there was no market other than military sales of any significance, the United States, despite its early lead, had fallen well behind the European powers in aeronautical technology.

The supply bureaus of the Army and Navy were accustomed to specifying the design of items they wished to purchase in all details, mess kits and warships alike. This was impossible for aircraft; nowhere in government did the necessary expertise exist. In 1915, after long and contentious debate, Congress passed legislation creating the National Advisory Committee for Aeronautics (NACA). The new organization, an operating agency despite its designation as a committee—NACA's main committee functioned as a board of directors and subcommittees provided technical guidance—was given the job of improving the relevant technology base in anticipation of

U.S. entry into the war underway in Europe. By the time NACA built and staffed a laboratory, World War I had been over for two years. While the agency went on in the 1920s and 1930s to make useful if modest contributions to aviation technology, conservative management and political opposition hobbled NACA throughout its existence, which ended in 1958 with absorption into the newly formed National Aeronautics and Space Administration (chapter 5).

From the beginning, NACA undertook R&D relevant to both military and civil aviation: the agency's creation marks the beginning of explicit dual-use policies.[18] Within the armed forces, however, research barely existed. During the interwar years, almost all the money available for military R&D—little enough, as table 1 makes plain—went for engineering, most of it quite routine. An occasional research project might be farmed out to NBS. Otherwise, the technical employees of the Army and Navy

Table 1   U.S. Military R&D, 1935–1945

	Millions of Dollars		
	Army and Navy[a]	Other[b]	Total
1935	$ 8.8	$ 0.8	$ 9.6
1936	12.7	1.2	13.9
1937	11.5	1.3	12.8
1938	13.9	1.3	15.2
1939	13.1	1.7	14.8
1940	26.4	2.2	28.6
1941	143.7	7.9	151.6
1942	211.1	16.0	227.1
1943	395.1	139.0	534.1
1944	448.1	835.2	1,283.3
1945	513.0	997.6	1,510.6

[a] Includes Army Air Forces.

[b] National Advisory Committee for Aeronautics (all years), Office of Scientific Research and Development (1941–1945), and Manhattan Project (1943–1945).

Source: 1935–1939—Army and Navy, Hearings before the Select Committee on Postwar Military Policy, House of Representatives, Seventy-Eighth Congress, Second Session, pursuant to H. Res. 465 (Washington, DC: U.S. Government Printing Office, 1945), pp. 228–229; National Advisory Committee for Aeronautics, Alex Roland, Model Research: The National Advisory Committee for Aeronautics, 1915–1958, NASA SP-4103, Vol. 2 (Washington, DC: National Aeronautics and Space Administration, 1985), Table C-2, p. 475. 1940–1945, all agencies—Federal Funds for Science XI: Fiscal Years 1961, 1962, and 1963, NSF 63–11 (Washington, DC: National Science Foundation, 1963), Table C-32, p. 136.

evaluated products offered by suppliers, whether aircraft engines or steam traps for the engine rooms of navy ships, suggested improvements, and occasionally channeled funds to those suppliers for developments they would otherwise not pursue.

What technological expertise the Army could claim resided mostly in the Corps of Engineers, the Ordnance Department, and the Signal Corps. Army engineers had run the West Point military academy until the late 1870s. Before the Civil War, the academy educated the majority of American engineers who could claim formal training of any sort. During the 1870s, with the Morrill Act providing grants of public lands to the states to encourage "agriculture and the mechanic arts" and technological practice beginning to shift from craft-based empiricism to the science-based methods that came to dominate over the next half-century, the number of American engineering schools rose from 7 to 70, the Army's role in technical training lapsed, and with it any strong impetus to keep up with technological change.[19] During the antebellum years, the Army's Ordnance Department also had a prominent place in the national technological enterprise. Responsible for both small arms and big guns, Ordnance employees made substantial contributions to the technologies of metalworking manufacturing. These largely ended with the Civil War. Demand for armaments greatly outstripped arsenal capacity and the North turned to private firms for something like half the number of weapons needed. After 1865, arms production fell to nearly nothing and the locus of innovation in manufacturing shifted to consumer goods such as sewing machines and bicycles. Given the growing military significance of telegraph and, in later years, radio, technical leadership in the Army, such as it was, passed to the Signal Corps, which gained an Aeronautical Division in 1907 and the following year took delivery of the Army's first airplane. The Navy bought its first planes three years later.

By this time, the Navy, like its counterparts in other nations, had become more technically competent, on an organization-wide basis, than the Army, a consequence of the shift from wood and sail to steam propulsion and steel-hulled construction. In the late nineteenth century navies were experiencing technological change so rapid that "it became literally true that a warship was obsolete before it was completed, in the sense that ships that were then already on the ways would easily be able to defeat it in battle."[20] William S. Sims captured the essence of contemporary naval technology when he wrote in 1909 to President Theodore Roosevelt: "Battleships are now great machines. A turret and its guns with their elevating and training gear, ammunition hoists, etc. is a mass of machinery. To handle them efficiently line officers must be competent engineers."[21]

Sims' words elided the divisions in labor and status between the Navy's "line" and engineering officers. Engineering officers, even if Naval Academy graduates, were ineligible for command; in the eyes of line officers—so named in recognition of what had been perhaps the premier test of skill and decisiveness during the age of sail, formation and station-keeping of warships in the battle lines called for by naval doctrine—they were necessary nonentities, supervisors of greasy machinery rather than fighting gentlemen. The line officers who ran the Navy showed little inclination to accept the advice of engineering officers on technical matters, much less solicit it.[22] Similar attitudes persisted even into the Cold War. The near-dictatorial control exercised by Admiral Hyman G. Rickover over the design of nuclear submarines generated much opposition within the Navy (chapter 6); that he was an engineering officer who would never have been permitted to command one of those submarines compounded the resentment (that Rickover, before transferring to the Navy's engineering branch in 1938, had briefly captained a minesweeper made no difference).

Before World War I, the Navy maintained somewhat closer working relationships with private firms than the Army. When government personnel in the Bureau of Construction and Repair laid down ship designs, for instance, they sometimes collaborated with employees of private yards.[23] The first ship of a new class would ordinarily be built in one of the Navy's own yards. Should government later choose to contract with private shipbuilders for additional construction, it knew what the job entailed and could compare bids against its own cost data.[24] The Army's Ordnance Department likewise drew on arsenal experience in negotiating contracts with private firms.

Army Ordnance was known for treating the rest of the Army almost as arbitrarily and unreasonably as it treated inventors and private firms.[25] The supply bureaus, their autonomy grounded in legislation, could and did ignore, circumvent, or defy the military chain of command up to and including the Army Chief of Staff and the Chief of Naval Operations, answering only to the Secretary of War or the Secretary of the Navy, and to them only in principle.[26] Not until the 1960s was their independence substantially curbed. With some exceptions for needs recognized as critical but for which the bureaus lacked internal capability, such as gunfire control (accurate targeting of fast-moving, maneuvering planes or ships depended on rapid solution of differential equations, hence on automation, a point that few in the military grasped before World War II), they showed little interest in the sorts of systematic searches for technological improvement common in private industry from the beginning of the twentieth century. Individual officers, whether or not attached to one of the bureaus, might be persuaded of the need for innovation, open-minded as to its sources, and responsive to private firms, but most of these men were in the lower ranks and dependent for influence on their ability to persuade superiors likely to be strong in their opinions but ignorant of technology (and perhaps resistant to logic). Rarely did the bureaus as organizations extend themselves to accommodate innovation, exchange information, or coordinate technical activities.[27]

Historians sometimes trace the roots of the so-called military-industrial complex to the Navy's transition during the nineteenth century from sail to steam and wood to steel.[28] The argument relies too heavily on the particular and quite specialized technologies of steelmaking and shipyard fabrication. It is certainly true that demand for armor plate hastened the shift from Bessemer steel to the nickel-alloyed products of open hearth furnaces, and that Bethlehem Iron Works and the Carnegie Steel Company and its successor, U.S. Steel, profited handsomely from sales for naval shipbuilding, thanks in part to their participation in an international cartel.[29] It is also true that the "New Navy" policy of the late nineteenth century signifies, in two respects, a shift in relations between government and industry. First, Congress directed the Navy to increase its purchases from private shipyards. Second, naval expansion marked the nation's first large-scale peacetime arms purchases. Yet the essential characteristic of defense acquisition since World War II is reliance on private firms for the *design* of complicated technological systems, not just production. Into the 1960s and 1970s, the Navy designed its own ships, even though private shipbuilders might take a hand in some tasks, whereas private firms from the very beginning designed aircraft for the Army and Navy. Neither service had meaningful capabilities in the infant technology of aviation, with World War I underway there was no time to develop such capabilities, and attempts to do so during the interwar years did not get far. In the transition from a system of government arsenals and shipyards to the privatized defense industry of today, aircraft led the way and shipbuilding lagged.

The Naval Research Laboratory (NRL) opened its doors in 1923, the first military facility with technical ambitions much beyond testing.[30] The laboratory, which at first employed only about 20 technical personnel and had a proportionately small budget, took over programs in radio communications and acoustical detection of submarines earlier scattered among naval stations and shipyards and in borrowed space at NBS. For some years, NRL managers were kept busy fighting off efforts to absorb the new facility into the Bureau of Engineering or otherwise turn it away from research. During the 1930s NRL more than demonstrated its value to the Navy through work on the radically new technology of radar. Yet even in

1940, with radar equipment designed by NRL (it was built by RCA) entering the fleet, conferring obvious and immediate tactical advantages by making both friends and foes visible in darkness or bad weather, the laboratory's budget was only $370,000. That same year, with German forces occupying Paris and bombing London, the U.S. government continued to channel more money to agricultural research ($29.1 million) than to military R&D ($28.6 million).[31]

Military R&D rose by more than 400 percent in 1941 and continued to increase through 1945. Existing R&D organizations exploded in size and new facilities sprang up as if overnight. During the war, Vannevar Bush's OSRD, staffed entirely by civilians, exerted influence out of all proportion to its share of funds, which was only a little over 7 percent (table 2). OSRD's leverage came from control over front-end research, low in cost compared to design and development of weapons slated for production.

An MIT engineering professor and administrator with strong convictions who became a close personal adviser to President Franklin D. Roosevelt, Bush could generally get his way even with recalcitrant generals (admirals, beginning with Chief of Naval Operations Ernest J. King, gave more trouble). He served on NACA's main committee from 1938 and as its chair in 1940–1941, coming to believe that existing agencies were too cumbersome and cautious to push rapidly into new technologies that could help win the war. For these tasks, Bush and his colleagues in OSRD tapped their peers among the nation's scientific and technical elite. Bush believed the military should stick to improving existing weapons, not try to develop new ones.[32]

OSRD channeled the bulk of its funds to research and prototyping contracts at newly established laboratories administered by universities.

Table 2   World War II R&D Spending

	Millions of Dollars[a]
Manhattan Project	$ 1,666
Army[b]	1,077
Navy	644
Office of Scientific Research and Development (OSRD)	270
National Advisory Committee for Aeronautics (NACA)	60
Total	$ 3,717

[a] Fiscal years 1941–1945.

[b] Includes Army Air Forces, which accounts for slightly more than half the total.

Source: *Federal Funds for Science XI: Fiscal Years 1961, 1962, and 1963*, NSF 63–11 (Washington, DC: National Science Foundation, 1963), Table C-32, p. 136.

Notable among these were the Radiation Laboratory at Bush's own school, MIT, which became the center of wartime radar research, and the Johns Hopkins Applied Physics Laboratory (APL), which developed the proximity fuze.[33] Bush and his staff worked hard to keep OSRD-funded programs tightly focused on weapons and to convince the military to move quickly to adopt and employ them. At the same time, Bush was content to steer clear of most issues concerning production. With a few exceptions such as proximity fuzes, which APL continued to oversee into production because they were so delicate and demanding to manufacture, OSRD and its contractors handed over technologies to the military when judged ready for engineering development. The services, in turn, contracted most of this work out to industry.

As the end of the war approached and OSRD prepared to shut down, the services, newly appreciative of R&D, began to put in place replacement structures of their own (chapter 5). Until the Korean War, funds were scarce and the scale small. The public was tired of war, wanted the troops home, an end to conscription and "foreign entanglements." With exceptions for the atomic bomb and jet propulsion, self-evidently too important to neglect, Congress and the Truman administration were disinclined to spend money on new weapons or R&D that might lead to them. While still busy with demobilization, the services found themselves grappling with the implications of nuclear weapons for roles and missions and drawn into national debate over universal military training. Economic reconversion proceeded rapidly, with less disruption and displacement than many had anticipated (even though employment in the aircraft industry, to take one example, fell from a wartime peak of more than 2 million in 1943 to 140,000 in 1946). For the military, by contrast, adjustment to peace and then to Cold War brought soul-searching, infighting, and distress.

The 1947 National Security Act granted the Air Force independence from the Army and placed what had been the War Department and now became the Department of the Army alongside the Air Force and Navy in a "National Military Establishment." Amendments two years later turned this into the Department of Defense. Far more upsetting to the leaders of the armed forces, the 1949 amendments ended their direct access to the president: service secretaries now reported to the Secretary of Defense. For the Navy especially, this was a wrenching change. Fearing loss of the Marine Corps to the Army and naval aviation to an independent air force (as in Britain), it had been fending off proposals for "unification" since the 1920s, when Army aviators began voicing proposals for tripartite organization of the nation's military forces.[34]

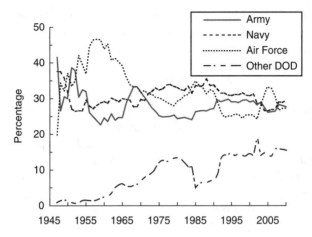

Figure 3   Service Shares of Defense Department Budget

Note: Navy includes Marine Corps. 2007–2010 estimated. Other DOD includes retirement pay through 1984, after which this item is included in service budgets.

Source: National Defense Budget Estimates for FY 2008 (Washington, DC: Office of the Under Secretary of Defense [Comptroller], March 2007), Table 6–3.

The Navy managed, in the legislative compromises of 1947 and 1949, to preserve nearly all the prerogatives of service autonomy it valued so highly. The creation of the Air Force nonetheless unbalanced accommodations between the Army and Navy that had been maintained for 150 years. While it now seemed possible for two of the services—i.e., the Army and its stepchild, the Air Force—to combine against the interests of the third, in the end "unification" had its greatest effects on budgetary matters. With a single appropriation replacing approvals by the congressional committees that earlier funded the Army (including the Army Air Forces) and Navy (including the Marine Corps) independently, budgetary competition now approached more closely to a zero-sum game, with the services fighting head-to-head for shares of overall Department of Defense (DOD) spending. As figure 3 shows, the Air Force, charged with strategic deterrence, the new national security imperative, and busy building up its fleet of long-range bombers and jet-propelled fighter-interceptors, was the big winner during the 1950s.

Gearing Up for Cold War

Well before the end of World War II, a number of high-ranking officials had come to believe that science and research would constitute the future basis of U.S. military power. In 1944, Under Secretary of War Robert Patterson announced that "We will make plans to suit our weapons, rather than weapons to suit our plans."[35] Patterson's statement might be taken to foreshadow nuclear strategy during the early years of East-West confrontation. More generally, however, his prescription could not have been applied: while R&D might tell planners what weapons it is *possible* to build, it cannot tell decision-makers what *should* be built. That is fundamentally a question of national security policy, and it would be some years before U.S. national security policy solidified in its postwar mold.

Stalin's intentions were unknown and U.S. intelligence agencies had little understanding during the second half of the 1940s of Soviet industrial, technological, and military capabilities. Much of the baseline information available to them, even to maps showing the locations of Russian cities, came from captured German files. The Red Army had fought a separate war against Hitler on a separate front, with no military-to-military cooperation to provide insight into the thinking of Soviet generals and their political masters. The Soviet economy remained as much a mystery to U.S. intelligence after the war as at the beginning; although the United States shipped large quantities of raw materials, machine tools, and military equipment to Russia, the Kremlin had consistently declined to explain why it wanted what it asked for.

High levels of uncertainty persisted through the 1950s. The United States put a great deal of effort into a profusion of weapons during the early Cold War years in part because the threat was so poorly understood. In the absence of knowledge of the technological capacities of the Soviet state, the Kremlin's geopolitical aims, and the warfighting doctrine of the Red Army, the U.S. military enjoyed great latitude in tailoring intelligence analyses to support the perspectives, preconceived notions, and more or less parochial interests of services and service branches. At one point, naval intelligence suggested that the USSR might be planning to build as many as 2,000 submarines.[36] Not until 1956, when U-2 spy planes started their overflights of Soviet-controlled territory, did more accurate assessments begin to become available. Early U-2 photos discredited claims of a bomber gap, revealing "fewer than three dozen of the new Soviet . . . (Bison) heavy bombers [at a time when] the U.S. Air Force was claiming that nearly 100 of the Bisons were already deployed."[37] Then in 1960 the CORONA series of reconnaissance satellites returned the first photographs of the Russian interior beyond the range of U-2 flights, filling in more gaps in U.S. intelligence.

During the years of greatest uncertainty, interservice rivalries in the United States were at a peak. As they struggled to fit nuclear weapons into their budgets, force structures, and doctrine, the Army, Navy, and Air

Force competed fiercely for control of aviation and rocket, missile, space-launch, and missile-defense programs. In attempting to shape definitions of roles and missions, each service found reasons to exaggerate particular dimensions of the Soviet threat and minimize others. Sometimes threat inflation reflected genuine uncertainty. Other times the reasons lay in budgetary politics or conflicts such as that between the Air Force and Navy over the merits of long-range, land-based bombers versus strike aircraft launched from the Navy's carriers for delivering nuclear weapons to targets inside Soviet Russia.

Despite gains over the years in baseline information, hindsight, aided by the opening of Kremlin archives, has revealed persistent overestimates of Soviet capabilities. As late as 1985, an ostensibly objective analyst could state, without qualification:

> Political leaders and defense decisionmakers agree that the Soviets have far exceeded the United States in military R&D . . . . Over the past 10 years the Soviets have spent in real terms some $185 billion more on military R&D than did the United States.[38]

Still, the exaggerations did not always go unrecognized:

> Why planners assumed the Soviet Union could launch simultaneous offensive thrusts in Scandinavia, Western Europe, South and Southeastern Europe, the Middle East, India, and the Far East is hard to fathom . . . .[39]

The 1950s are often recalled as a time when the United States chose to rely on its nuclear deterrent rather than spend large sums on conventional military forces. That is a bit misleading. With the reflexive fiscal conservatism it shared with the pre–Korean War Truman administration, the Eisenhower White House managed to hold down total military spending, in part through its announced policy of "massive retaliation." By brandishing the prospect of a nuclear response to almost any form of Soviet aggression, this policy was available to justify curbs on spending for nonnuclear weapons. Yet the always equivocal threat or promise of massive retaliation was widely disbelieved, both in the United States and abroad, and by the mid-1950s, with the U.S. military greatly strengthened as a result of the buildup spurred by the Korean War, the administration had begun backing away, if somewhat erratically, from the positions it had articulated earlier.[40] Eisenhower believed that any full-scale war with the Soviet Union would inevitably escalate to nuclear exchange. He also believed that so long as manned bombers, rather than ballistic missiles, remained the delivery vehicles for all or most of the warheads in the stockpiles of the United States and the Soviet Union—bombers that were vulnerable to air defenses—a nuclear exchange, no matter how devastating, would not end the war. In the extended conflict following, outcomes would be determined by residual industrial capacity and conventional military power.[41]

In the end, the Eisenhower administration's budget-balancing efforts had relatively little effect on the U.S. commitment to a high-technology military. American firms designed more new military planes in the 1950s than in all the years since—ten new supersonic fighters alone. By one account, aircraft R&D programs in the 1950s outnumbered those that followed during the 1960s and 1970s by 122 to 39.[42] The Pentagon put defense contractors to work on networked computers for command and control, new generations of highly sensitive radar equipment, ballistic missiles and early versions of cruise missiles, tactical missiles of all types and sizes.[43] DOD also funded fundamental research—in atmospheric physics, digital electronics, materials science (created as a freestanding discipline largely with Pentagon funds), the social sciences too. By studying, over many years, prospective designs for quiet submarines and techniques for finding and tracking enemy submarines in the deep ocean—work that ranged from basic research in oceanography to the development of towed sonic arrays—the United States gained knowledge for designing its own submarines and at the same time insurance against the possibility of a Soviet breakthrough in quiet submarines or in methods for tracking and targeting U.S. submarines. Despite the seemingly radical nature of some of the resulting innovations, most of the advances were incremental, as they had been with radar during World War II and as they continue to be today with most technologies, military and civilian alike.[44]

In retrospect, the 1950s seem a period of heroic technical ventures verging on the bizarre. Defense agencies channeled large sums to the Dyna-Soar space plane and nuclear-powered bomber mentioned in chapter 1. The Navy purchased a jet-propelled winged torpedo: dropped from a patrol plane safely distant from the enemy, the Petrel was to speed toward its target under radar guidance before descending by parachute to cover the last mile underwater. The Army, not to be left slogging in the mud, at one point proposed building tanks with detachable flying turrets. Some of what in hindsight appears to be overreach can be attributed to the unknown dimensions of the Soviet threat, some to overpromises made by the military or its contractors in their efforts to build political coalitions in support of this weapons system or that. Those bent on selling the nuclear-powered bomber, for example, had an easy time exploiting murky intelligence estimates as they extolled the merits of a plane able to cruise the skies indefinitely. With no need to refuel, nuclear-powered bombers promised a hedge against one of the chief worries of Air Force planners, a sneak attack that caught America's own bombers on the ground, where they were easily located and vulnerable, wiping out what was then the principal U.S. deterrent. Whether the Soviet Union had the capability to mount a mass attack on U.S. air bases was a question Air Force leaders were happy to leave unasked and unanswered.

CHAPTER 4

UNDERSTANDING ACQUISITION

By actual measurement in July 1985, legislation, regulations, and case law governing defense procurement alone cover 1,152 linear feet of shelf space.
— John F. Lehman, Jr., *Command of the Seas*[1]

In 1991, Secretary of Defense Richard Cheney canceled the Navy's $50-plus billion A-12 attack plane. After three years of full-scale development, the program had fallen a year behind schedule and was 50 percent over budget. The A-12 was hardly unique. The difference, as Cheney and others in the administration saw it, was that the nation could no longer afford to bail out undertakings that were in so much trouble. The deeper question is how to avoid such situations in the first place. To find answers, we must understand how the United States ended up with its current policies and practices. This chapter continues that story.

It begins by summarizing the primary acquisition issue facing the United States over the next two decades: a pipeline filled to overflowing with more weapons programs than the nation can afford (more accurately, more programs than it is likely to decide it can afford). Next the analysis turns to the historical roots of the dilemma, which go back to World War I when today's defense industry and the relationships that bind it to government began to take shape. At that time, Washington confronted a problem for which it was unprepared: buying planes that would not become flying coffins for the pilots of the Army's fledgling Air Service. With new generations of aircraft reaching the Western Front at six-month intervals—a faster pace than for integrated circuits today—the U.S. military, with hardly any capability in aviation, had no choice but to rely on private firms. Then and later, the government's preferred approach—purchase of experimental models that test pilots could fly, followed by competitive bidding for production contracts—failed to adequately compensate companies that won the "fly-off" but lost the follow-on procurement competition, and sometimes resulted in delivery of inferior or defective planes when a winning bidder failed to appreciate the engineering principles embodied in another firm's design. Some nine decades later, the underlying problem—how to manage development of systems as yet undetermined in function and form—continues to plague the Department of Defense (DOD).

What has been called the military-industrial complex originated in World War I and the interwar years. The label is misleading, certainly if taken to suggest alliance or intrigue for spending money on unneeded weapons. Many studies of acquisition have found "waste, fraud, and abuse" to be no more than minor concerns. At the same time, the tainted image of the military-industrial complex highlights one reason for the high costs, lengthy schedules, and uncertain outcomes of major defense programs: the tangle of laws and regulations that govern military contracting, many of these intended as deterrents, a consequence of fear of waste, fraud, and abuse.

The Acquisition Dilemma

In 1989, at the end of the Reagan defense buildup (which actually began in the last two years of the Carter administration), the defense budget approached $300 billion. With the end of the Cold War, spending did not exceed that mark until 2002, when the terrorist attacks of the preceding September and the fighting that followed in Afghanistan and Iraq spurred sharp increases. Even so, defense consumes only a little more than 4 percent of gross domestic product (GDP), far less than the 14 percent reached in 1953 during the Korean War or the 9-plus percent in 1968 during the Vietnam War (figure 4). Of course, the nation's GDP is much larger today and so is defense spending in absolute terms. With military expenditures falling in many countries, recent estimates put U.S. spending at around half the world total.

At the peak of the Reagan buildup in the second half of the 1980s, the defense sector employed nearly 7 million Americans: 2.2 million in the armed forces, 1 million civilian DOD personnel, and the remainder in industrial firms supplying the Pentagon. Since then, the ranks of the uniformed military have fallen to 1.4 million and DOD's civilian workforce to about 670,000.[2] Yet if not so large as before, DOD remains the biggest arm of the federal government by far, the biggest more-or-less unitary organization in the nation and the world. (Wal-Mart, the nation's and world's largest private employer, has about 1.3 million people on its U.S. payroll

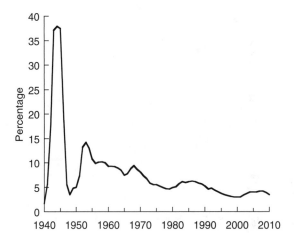

Figure 4   Defense as a Share of U.S. Gross Domestic Product

Note: 2007–2010 estimated.

Source: Historical Tables, Budget of the United States Government: Fiscal Year 2008 (Washington, DC: U.S. Government Printing Office, February 2007), Table 3–1.

and 1.9 million worldwide.) Size alone would make DOD enormously difficult to manage.

Each year Pentagon personnel execute over 5 million contract actions, as many as the rest of the federal government combined (by value, DOD contracts account for two-thirds of the total). In recent years, acquisition—i.e., RDT&E (research, development, test, and evaluation) plus procurement—has accounted for around one-third of DOD spending. (Most of the rest pays for wages and benefits and for operations and maintenance.) Acquisition dollars go primarily for what the military terms platforms (aircraft, ships, armored vehicles, satellites), systems (electronic countermeasures, command, control, communications, and intelligence), and weapons (guns, bombs, missiles). Platforms generally carry both weapons and electronics.

Over the past quarter-century, some 38 percent of acquisition funds have gone to the Air Force, 34 percent to the Navy (including the Marine Corps), and 18 percent to the Army.[3] Service shares have not changed greatly since the end of the Vietnam War. That is not true of the share of RDT&E relative to procurement, which climbed dramatically in the 1990s. With the end of the Cold War, procurement fell by nearly half while RDT&E remained roughly constant (in the range of $35 billion to $37 billion annually). The consequence, as figure 5 shows, was to push the ratio of RDT&E to procurement to unprecedented heights. Since the bulk

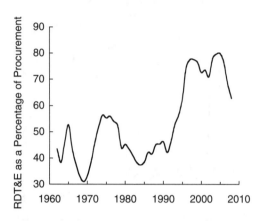

Figure 5   Military RDT&E Relative to Procurement

Note: 2007–2008 estimated.

Source: Historical Tables, Budget of the United States Government: Fiscal Year 2008 (Washington, DC: U.S. Government Printing Office, February 2007), Table 3.2.

of RDT&E funds, around four-fifths of the total, pay for the engineering development of weapons systems slated for eventual production, continuing high levels of RDT&E in the 1990s meant, quite simply, that the United States was continuing to develop new weapons much as it had during the Cold War. (Recent fluctuations reflect rapid rise in procurement spending, which has more than doubled since 2000.)

In the civilian economy, the ratio of R&D to sales is the usual measure of technological intensity. For U.S. industry as a whole, this figure has recently been in the vicinity of 3.5–4 percent. In high-technology sectors such as electronics and biotechnology, some firms spend more than 10 percent of revenues on R&D. But nowhere in the civilian economy does the percentage come anywhere near those in figure 5. (Start-up firms with little or no revenue may report high ratios of R&D to sales for a time, but cannot sustain them.) Such comparisons have limited significance for reasons including lack of correspondence between DOD's RDT&E categories and the accounting conventions used in industry. Still, they suggest the extent to which defense is sui generis, utterly different from any and all technical activities in the civilian economy.

The larger message of figure 5 is this: high levels of RDT&E spending in the 1990s reflect policy decisions by several administrations and Congresses to continue developing new generations of high-technology weapons much as if the Cold War still continued. Because it takes 10–15 years, sometimes more, to move a major program from preliminary design through engineering development and into production, much of the RDT&E spending of the 1990s went for the design and development of weapons and systems conceived in the 1980s to be produced in the early part of the twenty-first century. That describes the Air Force's F-22 as well as Navy ships "on the drawing boards and in the budget [that] are straight-line extensions of the Cold War design vectors for these ship types . . . ."[4]

If the basic thrust of acquisition policy did not change much over the Cold War and since—high technology weapons in profusion, many with overlapping capabilities—the quantities purchased have fallen as both costs and capabilities have increased. Many of the new weapons are far deadlier than their predecessors. Because they can do more, fewer are needed: a handful of precision munitions may take the place of hundreds or thousands of "dumb" bombs. Then too as weapons become more lethal, fewer platforms are needed for delivering them. In the early 1960s, the Air Force bomber fleet numbered in the thousands, including more than 700 B-52s alone. Today, the nation's heavy bomber fleet, in its entirety, consists of 21 B-2s, 67 B-1Bs, and 94 B-52Hs. (Only about half of these are combat-ready at a given time.) One reason why the heavy bomber fleet has become so small is that attack planes such as the stealthy F-117 and the Navy's F/A-18 carrying precision weapons can do far more damage than their predecessors. Although these planes have much shorter ranges and cannot carry the heaviest weapons (or as many lighter weapons), for some missions they now complement or replace heavy bombers. (The F-117, easy to see and target in the daylight unless flown at high altitude, is normally restricted to nighttime combat missions—but so is the B-2.)

Costs for weapons systems have been rising faster than the economy-wide rate of inflation for decades. Ever-higher price tags are a consequence of efforts by the military to cram as much "technology" as possible into their weapons systems in search of the greatest possible performance. The underlying tendency has been evident for many years and shows no sign of abating. In 1952 when the Navy refitted a dozen destroyers with the latest available radar and communications gear then available, the costs exceeded those for building and equipping those same ships, beginning with raw steel, less than a decade earlier. Over the period 1950–1980, inflation-adjusted costs for the electronics incorporated into fighter aircraft increased by 40–50 times, powerplant costs by 15–20 times, and airframe costs by a factor of 5.[5] Since the 1950s, the real costs of digital electronics have declined steeply and continuously in the civilian economy. The difference is that civilian applications are standardized and produced in volume, whereas military equipment is highly specialized and, with some exceptions, built by the handful (and sometimes by hand).

Rapidly rising costs were exactly the problem with which McNamara and the systems analysts he brought into the Pentagon struggled during the 1960s. The trends have only worsened since. The very demanding performance requirements imposed by the military mean years of heavy R&D spending—RDT&E for the F-35 Joint Strike Fighter (JSF) is expected to reach $45 billion before production starts. Many years pass in efforts to push the state-of-the-art ahead. Only then can detailed design even begin. Pressing so hard on the technological frontiers leads to systems that are very costly to build, while small production quantities reduce opportunities for economies of scale and learning (although these are minor factors in cost escalation by comparison with the other factors at work).

In part because military equipment is too expensive for routine replacement, it must then remain in service much longer than typical consumer products and many capital goods. Lifetimes range from perhaps 20 years for tactical missiles to 50 years or more for some ships and planes. Boeing began design work on the B-52 in 1946. Those B-52Hs still in service incorporate many modifications, including new wings and engines. Even so, upgrades, with exceptions mostly for electronics, are restricted by the original design architecture—the overall configuration of the system. A B-52 is still a B-52. There is no way to make it stealthy, for example (the chief advance in bombers since the original design was laid down). Thus when a new program begins, the military, knowing the system will remain in its inventories for decades, pushes for capabilities that will remain "state-of-the-art" well into the future. This in itself forecloses prospects for controlling costs and schedules. No matter the effort expended to press ahead technologically, systems take so much time to develop that they almost inevitably incorporate outdated components and subsystems by the time they reach the field. The F-22, for example, included "200 obsolete military-specified electronic parts" at the beginning of production, even though the plane's "avionics systems [had] undergone four technology refresh cycles."[6] The dynamic is perverse and self-defeating:

> In developing its weapons, the Navy's goal frequently is to build a state-of-the-art system on the cutting edge of technology, rather than a simpler system that is good enough to get the job done. The result is that developmental work is always halfway finished, and its goals are never reached.[7]

Once systems do reach the field, it is always possible that potential adversaries will be able to find—or purchase—countertechnologies.

F-22 proponents built their argument for the plane around two main themes: the ability to cruise at supersonic speeds without the fuel consumption penalty of using afterburners ("supercruise"), and a radar cross-section said to be that of a gumdrop. A great deal of work, in the United States and abroad, has gone into radar and infrared technologies for overcoming stealth (in part because stealthy cruise missiles pose grave threats, especially to ships). As detection methods improve, they will almost certainly leak out to any state (or terrorist group) willing and able to pay the price. That price may be high in absolute terms, but it is likely to be low relative to the costs of stealthy aircraft. The predictable result, as in any arms race: ever-escalating expenditures in vain efforts to stay ahead.

The wish lists of the services are perennially long and costly. In 2002, when officials began to float proposals to reduce the F-22 procurement to 180 planes, the response of the Air Force general serving as "program executive" for fighters and bombers was to declare: "Conceptually, the Air Force still thinks we need the 762 airplanes [originally planned]."[8] The Navy is buying more than 450 F/A-18E/Fs, a major redesign of earlier F/A-18 models, at a program cost expected to exceed $50 billion, and is scheduled to begin receiving the multiservice JSF around 2010. The JSF program is projected to absorb some $225 billion, a sum that will almost certainly increase unless production plans, already cut from nearly 2,900 planes to fewer than 2,500, are further reduced. The Navy commissioned the first of a projected 30 Virginia-class submarines in 2004 and wants new classes of cruisers, the as yet unnamed CG(X), and destroyers, the DDG 1000 Zumwalt, as well as the Freedom class of Littoral Combat Ships and a new generation of nuclear-powered aircraft carriers, the CVN 21. Expenditures for some 400 V-22 Osprey tilt-rotor aircraft, a Marine Corps program clouded for years by technical problems and a series of crashes, will come to about $50 billion. The Army has begun development of its Future Combat System (FCS), intended to be "overwhelmingly lethal, strategically deployable, self-sustaining and highly survivable in combat through the use of an ensemble of manned and unmanned ground and air platforms."[9] Foot soldiers will be linked into the overall system via wireless computing. Recent estimates by the Office of the Secretary of Defense (OSD) suggest that acquisition costs may reach three times the Army's 2003 figure of $77 billion.[10] Dozens of less visible programs, including electronics systems and guided missiles of many types, along with updates to existing equipment, add many more billions to the acquisition budget.

Future spending on systems already in production, such as the F/A-18E/F, can be estimated with reasonable accuracy. Systems in early-stage development almost always end up costing far more than the services say they will. The Navy puts the cost of each of the first two DDG-1000 Zumwalt destroyers at $3.0 billion; the Congressional Budget Office (CBO) expects these ships, stealthy, highly automated, and designed, like the Virginia, for attacking targets on land, to cost over half again as much, or $4.8 billion.[11]

For years, analysts inside the Pentagon and outside have warned that buying most or all of the systems in various stages of planning and development would require large sustained spending increases. CBO suggests that annual procurement budgets in the range of $140 billion over the period 2012–2024 would be needed.[12] That is well above 2007 procurement outlays of about $100 billion, as well as DOD's latest projections, which reach $115 billion in 2012, while there can be no guarantees that Congress will appropriate (or the White House request) even that much.[13]

Although the A-12, and more recently the Crusader and Comanche, were canceled outright, that is unusual. When there is not enough money to go around, the usual response of Congress and the administration is to cut production, as for the F-22. Because RDT&E may account for one-third or more of program spending and cannot be reduced unless the program is fundamentally restructured, when spread, along with other fixed costs, such as tooling and equipment, over fewer units, the costs allocated to each unit rise sharply, as again illustrated by the F-22 (and B-2).

Too many programs chased too few dollars throughout the Cold War. The larger difficulty is that it seems almost impossible to get the services to think and plan in terms of what they actually need rather than what they want.

The Heritage of Two World Wars

In recent years, some 600 laws and thousands of pages of regulations have governed DOD acquisition. This baroque structure of policy and practice has its origin in World War I, with still earlier antecedents.

Aviation: Infant High Technology

During World War I, procedures geared to buying boots and rifles proved woefully inadequate for equipping what was then the Army Air Service. With the nascent technology of aviation evolving rapidly in the hot house of combat in Europe, no one in government had the expertise to prepare detailed specifications as a basis for procurement competition based on sealed bids. All the Army or Navy could do was ask their pilots, junior officers with no combat experience and little if any technical training, to fly such planes as might be available and give their opinions.

Few if any innovations of World War I compared in rate of technological advance to aviation. Poison gases came from an established, science-based industry populated by chemical firms that had long supplied militaries with gunpowder and high explosives. Tanks, though powered like aircraft by engines based on automotive practice, were crude contrivances having more in common with agricultural machinery than with planes. Submarines, like all warships, drew on a naval technology base dependent on empiricism rather than science and theory.[14] The knowledge base for aircraft included both craft traditions and aeronautical science, but engineering knowledge was rudimentary and the science base had barely begun to develop.

Aeronautical design could not proceed as other military technologies then did. The traditional engineering approach worked for engines, but not for airframes. (Even today, theory makes relatively few contributions to the design of piston engines because of the complexity associated with "unsteady" flows of fluids and heat; advances result primarily from trial-and-error, intuition, and empiricism.) The aerodynamic forces that kept planes aloft and enabled pilots to control them remained a mystery: theory lagged behind practice at a time when aviation was caught between its bicycle-shop origins and a research base that had hardly begun to emerge.[15] Lack of accepted theoretical principles and analytical methods elevated the importance of tacit know-how, unwritten and unarticulated, yet trial and error was too dangerous a recipe. Aviation differed from earlier science-based innovations such as telegraph and wireless, dependent on electrical and magnetic phenomena no more self-evident or visible than those of aerodynamic lift and drag but reasonably safe subjects for experiment by comparison. In aviation, trials based on guesswork and intuition had failed again and again. The Wright brothers succeeded through painstaking experimental work tightly focused on problems personally encountered in flying gliders and then powered craft of their own design. If popularly remembered as bicycle mechanics, the Wrights were two of the most capable and creative engineers of their time.

No more than about 300 planes had been built in the United States between the Wright brothers' first flight and the beginning of the war in Europe. As the United States edged nearer to entry, hardly anyone in official Washington could bring informed technical judgment to bear on aircraft purchases. Neither Army nor Navy was accustomed to tapping outside sources of advice; if they wanted help, civilian scientists or engineers might be requested to volunteer their services (or in wartime drafted, with commissions for more senior figures). In this case, there was hardly anyone to ask.[16]

As we saw in the preceding chapter, procurement of technology-intensive matériel such as warships and armaments ordinarily began with design and development in government facilities, followed by production sufficient to verify performance and determine costs. This approach could not work for aircraft, which only a handful of small companies knew how to design and build. They had a monopoly on know-how but limited resources other-wise. Unlike arms-makers and shipbuilders, moreover, they were unfamiliar with the ways of government and lacked the political and establishment connections of the much larger automobile and auto parts firms that inserted themselves deeply into aircraft production during 1917–1918.

Aircraft Design: Performance versus Production

Embarking on futile attempts to mass produce aircraft during World War I, the U.S. government made two related mistakes.[17] The first was failure to appreciate the need for frequent design changes as a result of learning, both technological and doctrinal, in far-away Europe. This had implications both for design and for continuity in production. The second mistake was failure to grasp the necessarily craft nature of airframe fabrication at relatively low levels of output. With skilled workers and hand methods, design modifications could be incorporated easily. Even if designs could have been standardized, which was impractical because of the rate of technological advance, mass production as exemplified by automobile assembly lines would have been impossible. U.S. factories turned out about 14,000 planes (and 42,000 engines) over the 19-month period ending with the Armistice. These were prodigious figures relative to past output. But they did not approach the totals for the European powers, which built something over 175,000 planes in total, much less mass production in consumer goods industries: by themselves, Henry Ford's factories turned out nearly 400,000 Model Ts in 1915. For aircraft, it made no sense to try to replace skilled craftsmen and general-purpose tooling with specialized equipment operated by unskilled or semiskilled workers—the mass production model—even if aircraft designs had been stable, which they were not.

Continuous innovation driven by combat experience meant month-by-month design changes. Holley summarizes:

> Continuing superiority requires continual change. Every innovation intro-duced by the enemy must be outmatched. Superior performance in aircraft is the sum total of many components—range, speed, climb, maneuverability,

fire power, and the like—each conditioned by thousands of features of design: here a change in engine cowling to improve cooling and increase horsepower, there a better gun mount to enlarge the field of fire, and so on in an endless procession suggested by experience in the field and innovations on the drawing board. Fluidity of design is a requisite for superior weapons. Mass production, on the other hand, lies at the opposite extreme.

To freeze design is to facilitate production. The economies that stem from bulk purchase, long production runs without retooling, and the wider use of jigs and fixtures that permit more extensive employment of semiskilled labor, all require stability of design . . . .[18]

Detailed specifications could not be written as a basis for soliciting bids. Efforts to do so resulted in the U.S. Army buying aircraft (including several of foreign design) that were outdated before delivery. If flown in combat, they would have been literal death traps. Indeed, the British-designed De Haviland DH-4, produced by the thousands in the United States, came to be reviled as just that.[19] The successful aircraft of World War I came from producers on the continent and in Britain able to incorporate the lessons of combat quickly and content to rely on craft-based fabrication techniques.

In the end, fewer than 1,000 U.S.-built aircraft reached forward areas in Europe. None saw battle. American pilots flew French and British planes (just as Americans manned artillery provided by their allies). Even so, aviation had begun to alter the U.S. government's approach to procurement. After much conflict and controversy, cost-reimbursable contracts (e.g., cost plus fixed fee) became an acceptable alternative to fixed-price contracts based on sealed bids. Cost accounting, a primitive discipline at the time, became one of the consuming issues of procurement.

A quarter-century later, aeronautical technology had matured to the point that design standardization was practical and American industry had sufficient time to put its enormous productive capacity to work. During the 1920s and especially the 1930s, a host of innovations more radical than those seen during World War I transformed aviation. While the theoretical base remained thin, design practices became steadily more dependent on engineering science as all-aluminum low-wing streamlined monoplanes replaced wood, wire, and fabric biplanes.[20] The essence of engineering science, in aeronautics as in other fields, is the parallel advance of theoretically-based mathematical modeling and empirical studies that validate, refine, and extend those models (e.g., by pinning down adjustable parameters and probing limits of applicability). The origins lie in the eighteenth and nineteenth centuries, when analytical methods based on higher mathematics began to be developed to address pressing practical problems such as the strength of iron and steel construction. Craft knowledge provided little guidance for designing structures made of these new and unfamiliar materials. Bridges fell down and steam-engine boilers exploded. (So did cannon like the monstrous "Peacemaker," which blew up in 1844 during a demonstration on the Potomac River, killing six people including the secretaries of State and the Navy.) Calculus-based methods of applied mechanics could be applied to the analysis and design of aircraft structures (although these methods were not necessarily known to early designers). Aerodynamics differed. Flows of air were invisible; unlike the water moving around a ship's hull, nothing could be seen of the eddies and turbulence that contributed to sudden loss of pilot control. Lift was a puzzle; so were the sources of stability in flight. Even so, research-based methods began to supplement and eventually supplant the enlightened empiricism of aviation's pioneers. Theoretically based understanding, exemplified by the work of Theodore von Kármán, a German émigré who joined the Caltech faculty and later became a principal technical advisor to the Army Air Forces, contributed to reductions in drag and advances in stability and control that combined with more powerful engines fueled by high octane gasoline to give steady increases in performance as measured by speed, range, rate of climb, payload, and ceiling—parameters of even greater importance to the military than the commercial carriers that had begun to fly passengers and mail.

By the end of the 1930s, piston-engined planes were bumping against technological limits. Designs were not static, but continuing increases in performance, for fighter-interceptors especially, would require the radical innovation of jet propulsion rather than incremental advances of the sort that remained possible with piston engines and propellers. The new technological cycle gained momentum only in the 1950s. By comparison with World War I, then, World War II took place during a period of comparative stability in aviation technology. Standardization for volume production made sense. In 1940, President Franklin D. Roosevelt called on American firms to ramp up production capacity to 50,000 planes per year, an order-of-magnitude increase. An unrealistic target in the near term, Roosevelt was sending an unmistakable signal: industry should plan for high-volume production and the Army and Navy should begin training pilots, crews, and mechanics by the tens of thousands.

As war approached, the major powers had to decide what kinds of planes to build. Bombers like the B-17, able to hopscotch the Atlantic via Canada and Greenland to bases in England from which they could reach Germany, came from aircraft companies that had been working on passenger planes able to traverse the distances separating major North American cities. Firms such as Douglas, before the introduction of its mainstay DC-3, had earned much of their revenue from military sales, U.S. and foreign, putting the cash and know-how generated toward the design of commercial aircraft. Air carriers, meanwhile, had begun to earn substantial sums flying the mail, thanks to subsidies from the post office. Their revenues growing, airlines ordered larger, more powerful planes. When the Army conducted a design competition for long-range heavy bombers in 1933, Douglas drew on its commercial experience base for a twin-engined bomber proposal. Boeing, a small firm that had specialized in fighters during the 1920s, presented a concept that became the B-17. Once it had demonstrated with prototypes for the B-17 that four engines could be integrated into a bomber, this became the starting point for the Stratoliner, which entered commercial service in 1940. Boeing won further Army contracts for the B-29, on the drawing boards by the time of Pearl Harbor (Convair's design studies for the much larger intercontinental B-36 had also begun), grew enormously during the war, and went on to design jet-propelled bombers, transports, and tankers for the postwar Air Force and commercial jets beginning with the 707.

Planes such as the B-17 could not be built without engines capable of producing high power at high altitude, where thin air reduced both aerodynamic drag and the output of piston engines even when festooned with superchargers, intercoolers, and water-alcohol injection. Demand for long-range transport planes in Germany, geographically compact compared to the United States, had been limited. Although comfortably ahead in aerodynamics, Germany tried without success to develop long-range heavy bombers. Her industrial firms were unable to design and produce the powerful and reliable engines needed.[21] During World War II, the United States and Britain built a total of 130,000 medium and heavy bombers. These dropped some 2 million tons of high explosives on occupied Europe and Germany. Hitler's factories turned out but 17,500 bombers, the great majority of which had payloads perhaps one-quarter of those of American and British heavy bombers. The Luftwaffe managed to deliver fewer than 75,000 tons of bombs to targets in Britain.[22]

While Japan built 53 different models of aircraft for its navy and 37 for its army, the United States concentrated on production of 18 main types for Army, Navy, and Marine Corps combined. American factories turned out many other planes in smaller numbers, but these 18 models accounted for nearly 90 percent of the 275,000 military aircraft produced from 1939 through 1944, a total that exceeded the combined output of Germany (where Hitler did not put the aircraft industry on a maximum production footing until 1944) and Japan by nearly 100,000. British production added about 120,000 to the Allied total, and the Soviet Union, as noted in the preceding chapter, another 137,000. Large as these totals appear, even in the United States they did not approach mass production levels: at the peak in 1944, 8,000 planes rolled out of American plants in a single month—an impressive figure, yet only a half-day's output for the auto industry.

Design changes were continual, but mostly at detail levels; basic design architectures did not change. Standardization eased the task of keeping logistical pipelines filled. Mechanics worked on the same planes day after day and could devise short-cuts for maintenance and repair. Pilots could be trained in mass to fly P-47s, B-17s, B-25s. If any nation has had a surplus of capable military pilots, it was the United States during World War II. Japan trained 5,400 pilots in 1943, the United States nearly 83,000. American air-crews could be rotated home to convey the lessons of combat to trainees still in flight school; German and Japanese pilots flew until they died.

This time Washington made the proper choice when it opted for large-scale production of standardized designs. But the Army did not select the right mix of planes. The National Defense Act of 1920 had given the Air Service, alone among combat branches, full control over its own procurements. Following World War I, aviators struggled just to get a hearing from Army leaders. Then in the 1930s the high command walled off the Air Corps, leaving it with "unformalized autonomy."[23] Once free to choose the planes they wanted, Army aviators not only bought bombers in preference to fighters, they declined to purchase long-range escorts to accompany the bombers to their targets. Doctrine held them to be unnecessary. Planes like the B-17 Flying Fortress would be able to protect themselves. In the event, they could not do so.

Tanks

Technological change on the ground proceeded slowly compared to that in the air; standardization was never problematic for technical reasons. Nonetheless, during the interwar years the Army failed to settle on accept-able designs for either trucks, the responsibility of the Quartermaster Bureau, or tanks, that of the Ordnance Department. Weakly situated within the Army, the Quartermaster Bureau could not persuade the branches that used trucks to agree on common types.[24] The far more pow-erful Ordnance Department tried to dictate the design of tanks as it dictated the design of guns. While the Army managed to rationalize truck design and production by the late 1930s, because of Ordnance's recalcitrance the United States entered World War II with few and inferior tanks.

Accustomed to doing things its own way, the Ordnance Department disregarded the wishes of the combat arms. Some of its prototypes flatly contravened guidelines set down by the Tank Corps, such as rear-mounted engines to reduce vulnerability to enemy fire.[25] Because Ordnance would not pursue design concepts that the Tank Corps might accept, and because there was so little money for new equipment, French-designed tanks of World War I vintage (some built in U.S. factories) made up much of the Army's inventory. In 1939, the U.S. Army could field no more than 450 tanks—less than half the number of Polish tanks so easily rolled up by the German army that same year.[26]

The Ordnance Department could frustrate the rest of the Army in part because combat experience during World War I had been insufficient to clarify the place of the tank in military doctrine. As late as the Spanish civil war, there were no widely agreed templates for armored forces. Not until World War II was well underway did it become plain that artillery would be the best antitank weapons, followed by other tanks, and that the specialized "tank destroyers" carrying bigger guns but less protective armor built by a number of countries between the wars were of little use.

As the primary innovators, the British had taken more lessons from armored combat in World War I than other nations. Even so, they struggled inconclusively during the interwar years with planning questions. At least those questions got an airing. The U.S. Army, unlikely as it may seem in retrospect, defined its primary mission as waging defensive battles on the North American continent, where tanks would conduct "cavalry-like rampages in hostile rear areas."[27] Because no enemy could expect to succeed in landing heavy tanks in large numbers on America's beaches for U.S. armor to oppose, because the country was so large, and because the National Defense Act of 1920 had reduced the Tank Corps to an adjunct of the infantry (and because of the continuing mystique of horse cavalry), mobility headed the Army's list of tank design requirements. With the Tank Corps absorbed into the infantry, officers with World War I armor experience who hoped to advance had little choice but to switch their allegiance. The chief of the infantry bluntly stated his expectations in 1928: "The tank is a weapon and as such it is an auxiliary to the Infantryman, as is every other arm or weapon that exists."[28] Voices that might have favored mechanization, not only purchases of more tanks but of motorized artillery and troop carriers, were submerged or silenced. Planners gave little thought to doctrinal questions debated in England in the daily papers as well as the War Office, debates that may have been facile but at least took place.[29] Compounding their error, infantry leaders failed to stand up to the Corps of Engineers, which for a number of years refused flatly to supply portable bridging units able to support more than 15 tons (Germany's Tiger tanks weighed 55 tons).

Domestic politics also shaped the Army's choices. The high command did not wish to be seen preparing for another war in Europe, the sort of war that would call for heavy armor.[30] By restricting itself to light tanks, the Army could affirm to isolationists inside and outside government, who had seen to it that as late as 1939 the United States had fewer men under arms than Switzerland, and especially to congressional appropriators, that American soldiers would not again be sent to fight and die on European battlefields. For such reasons the United States entered World War II with tanks that could not stand up to German armor. Even at the end of the war, U.S. tanks were inferior. In neglecting tanks, the Army also neglected anti-tank weapons, sending troops overseas without adequate means to protect themselves.

Not until the fall of France in 1940 did the U.S. high command accept the need for mechanized forces. At that point, the Army simply copied the structure of Germany's Panzer divisions.[31] Tanks now had to be built in a hurry, and that was something the United States could accomplish. If American tanks could not match those of Germany in fighting power, American industry could turn them out in unequaled numbers—25,000 in 1942, compared with 330 two years earlier.

Costs and Competition: Who Pays for Development?

About 50 U.S. companies built planes during World War II.[32] At any one time, each of these firms might have 300,000 purchase orders outstanding with subcontractors and suppliers. Some 30,000 unique parts went into a B-24. Growing technological complexity brought greater administrative complexity. While major design changes were few, continued flux at detail levels generated blizzards of engineering change orders, required armies of clerks to track and record modifications, monitor orders and inventories, impute costs, track and coordinate spare parts inventories and the work conducted in modification centers remote from final assembly plants.

It proved impossible to estimate costs with any certitude in advance of production. Sometimes bidders erred on the low side, losing money on fixed-price contracts. Other times costs proved lower than projected, lead-ing to profits viewed by Congress and the public as exorbitant. The alter-native to fixed-price contracts was to reimburse firms for their actual expenses, with an added margin for profit. Such arrangements became widespread even though it was difficult to structure incentives to reward efficiency.[33]

Cost-reimbursable contracts posed many accounting issues. Should government pay the full cost of equipment and tooling that might be converted to commercial production after the war? What about firms that doubled or tripled executive salaries once war production began?[34] Cost-based contracts kept legions of bookkeepers, accountants, and auditors at work compiling and investigating allowable charges. One aircraft plant was able to lay off 1,800 clerks, bookkeepers, and accountants after a shift from cost-reimbursable to fixed-price contracts.

Firms that made planes and powerplants viewed design and development as a lead-in to production. They expected profits to follow once design and development gave way to procurement. The firm that designed a proto-type accepted by the Army or Navy generally got the production contract. But not always. Congress has continually pushed for "competition," despite the lack of a functioning market, arguing that by decoupling design from production and contracting with the low bidder, government could save the taxpayers' money. Defense firms, for their part, held that their designs, which drew on deep reservoirs of know-how, were proprietary. Without the expectation of subsequent procurement contracts, they had no reason to devote their best efforts and most talented engineers to mili-tary work, or indeed to undertake such projects at all.

The services generally sided with their suppliers in the belief that negotiated cost-plus contracts and close working relationships with industry were most likely to result in the equipment they wanted.[35] But again, not always. The Navy's Bureau of Ordnance (BuOrd), for example, although it worked closely with several private firms, "did not see research and development as a specific activity. When the navy wanted a new machine [e.g., for gun control], it simply specified one and ordered it from a com-pany. BuOrd then did what it pleased with the technology, including awarding production to other firms."[36]

During World War I, Congress held, in essence, that aircraft should be purchased like other matériel. Rifles were put out for competitive bids. Why not planes? The congressional position was naively simple. The government bought prototypes for its test pilots to try out. The govern-ment owned these prototypes and, in the congressional view, by extension the rights to their designs. Once the Army or Navy selected a plane, any and all comers should be allowed to bid for the production contract. That was "free enterprise." The more technically competent aircraft companies, those with engineering staffs as well as manufacturing capacity, countered that sale of one or a handful of prototypes did not constitute sale of intel-lectual property and could not be expected to fully and fairly compensate them for the true costs of design and development. They argued not only that they retained the rights to their designs, but that other bidders could not possibly do a satisfactory job of reproducing those designs. Prototypes were commonly delivered without drawings, much less technical calculations and data gathered in experimental facilities such as wind tunnels. Design changes, moreover, were continual, a normal part of development—indeed its essence—and might not always be documented fully and carefully. Because prototypes embodied tacit know-how and company-specific practices, it was inconceivable that another firm, lacking even drawings, could prepare a proper bid. It would only be guessing, might well underbid and then be unable to deliver on its promises. To be sure, a low-bidding firm would have to create or re-create a full set of drawings, process sheets, and bills of materials before beginning production. But the relatively mechanical exercise of tearing apart a prototype and taking dimensions off thousands of parts hardly amounted to "reverse engineering." That required higher-level skills, which the low-bidding firm might or might not possess. Even if it did, its engineers could not possibly understand the full range of technical decisions that had gone into the original design, either conceptually or at the detail level, and therefore would be poorly positioned to incorporate the further modifications that would surely be necessary as flight time began to accumulate. Although it might be necessary during a war for other firms to build entire aircraft or major subassemblies, the company that had laid down the original design should retain control over engineering and take the lead in modifications. The Army or Navy would want improvements, deserved them, and only in this way could be sure of getting them.

From the beginning, experience with cost-based competitive bidding proved these views correct. Delays were the norm as low bidders struggled to replicate prototypes designed by others. On occasion, the winning bidder ended up redesigning an entire plane, in which case it lost money and the government got what in effect was a different aircraft than it had ordered. In an egregious case in 1919, the Army took delivery of 50 Orenco-D pursuits—designed by Ordnance Engineering Company but built by Curtiss—then had to destroy them as unsafe to fly. By then, Ordnance Engineering had gone out of business.

Today as in the past, defense firms are reluctant to incorporate proprietary know-how in proposals to DOD if they fear loss of control over "knowledge capital." Low bidders sometimes find themselves redoing another company's design simply to understand how to build it. As they work through the difficulties, costs rise; the low bidder may seek reimbursement for at least some of these "unanticipated" expenditures—and, given the rules and regulations governing procurement, may well prevail.

Regulating Procurement: "Waste, Fraud, and Abuse"

Quite rightly, Congress and the public insist on close and constant oversight of government purchases in the name of accountability. The rules and regulations do a reasonable job, most of the time, of deterring outright fraud and exposing malfeasance when it occurs. They do nothing to address the technological reasons for poorly performing weapons systems. And rather than deterring waste, the rules contribute to it by driving up costs for compliance and resolution of disputes. Some contested claims from World War I remained before the courts during the World War II. By the late 1930s, "Every statute upon the books was encrusted with an intricate overlay of judicial decisions, Judge Advocate General and Attorney General opinions, and Comptroller General rulings as vital to the procurement process as the statute itself."[37]

Conflicts over acquisition law and policy between the executive and legislative branches and between industry and government have two fundamental sources: limits on the discretion of government employees intended to curb favoritism or worse and insistence on competition open to all comers. Jealous of its prerogatives and on guard against expansion of power in the executive branch, Congress has never been willing to accept that DOD and the services could manage acquisition honestly, efficiently, and equitably. All too aware from personal experience of the temptations of power, Congress insists on tight controls to limit opportunism by public officials. Not a few members have shared the impression, widespread among the public at large, of an unholy alliance between DOD and its contractors—a military-industrial complex in which billion-dollar deals and $2000 coffee makers alike get little scrutiny by outside parties and cost-reimbursable sole-source contracts serve as open invitations to fraud and abuse.[38]

Allegations of corruption involving public officials, defense firms, and, if often only by implication, the military (uniforms make risky targets for flag-waving politicians) reached a peak in the mid-1930s, when a special committee of the Senate chaired by Gerald P. Nye held a series of hearings on war profiteering and the international arms trade. In investigating the practices of shipbuilders and munitions suppliers, financiers, and the Du Pont company, the Nye committee was responding to, and furthering, belief that an international conspiracy of arms dealers instigated wars large (World War I) and small (the conflict then underway between Bolivia and Paraguay), earning unconscionable profits by selling to all sides. Senator Nye and his colleagues, lacking the power to probe dealings by foreign firms on foreign soil, had little chance of uncovering evidence of an international "ring" of arms suppliers. The committee did manage to demonstrate that American firms paid frequent bribes overseas, particularly in the Far East, that U.S. shipbuilders had probably rigged bids for U.S. Navy contracts, perhaps with the tacit or active approval of naval officers, and that the profits of American munitions suppliers on sales to the U.S. government had sometimes been large compared to other lines of business. That these were meager findings after nearly 100 hearings did little to disabuse conspiracy theorists who preferred to believe the "merchants of death" too clever to be caught.[39]

Postwar studies of defense procurement have generally found fraud and abuse, though not waste, to be minor concerns.[40] Scandals do occur, like that revealed in 2004 when a long-serving Principal Deputy Assistant Secretary of the Air Force pled guilty to steering contracts to Boeing before leaving to work for that firm. But their frequency is low, if only because the great majority of those in position to be tempted realize the chances of eventual exposure are high.

Congressional insistence on competition open to all comers has a comparably raw political source, as a response to constituent expectations. There are always firms that want a piece of the government's business and elected officials ready to oblige them. Yet it would be wrong to presume that members of Congress insert themselves into procurement decisions chiefly because of the pork barrel. If, on the one hand, members routinely try to steer contracts to their districts, on the other hand most senators and representatives, most of the time, view national security as among their graver responsibilities. While few members (or staff) have a deep grasp of military affairs, and fewer today than during the Cold War when there were more veterans in Washington and more members of Congress prepared to give national security sustained attention, political manipulation of major defense programs seems to have been less common than critics sometimes allege.[41] Congress is always making decisions on issues that members cannot expect to fully understand, whether the long-term future of social security or prospects for ballistic missile defense: that is the nature of the body. In this light, complaints of congressional meddling are misplaced. The most fractious part of a government disorderly by design, Congress meddles in everything. When it comes to defense, the weightier objections have to do with neglect of serious issues even as committees, subcommittees, and individual members preoccupy themselves with trivia to score political points.

As Secretary of Commerce and later as President, Herbert Hoover sought to rein in political forces and run government more like a business. In the 1960s, Robert McNamara attempted to bring contemporary management methods to the Pentagon. Not a few of his reforms were based on ideas that had been circulating in Washington for years; in considerable measure, McNamara's accomplishment consisted in forcing through their implementation. The 1984 Competition in Contracting Act has fostered use of commercial off-the shelf (COTS) parts and equipment, widely available at known prices, in military systems, so that DOD can take advantage of market forces acting in the civilian economy. Since the 1990s, OSD has put in place a further series of reforms aimed at "Adopting Best Business Practices."[42] DOD-specific practices and procedures raise procurement costs by perhaps 15–20 percent over otherwise comparable private-sector transactions.[43] Reducing this by even a few percentage points could save billions of dollars annually. Sometimes only a matter of paperwork—e.g., certifying that the price to the government does not exceed the lowest price charged to other customers—in other cases military standards and specifications ("milspecs") mandate specialized functional or quality tests that drive costs sharply upward even for mass-produced COTS items that differ only in labeling and certification when purchased for DOD.[44]

Public-sector procurements will always attract bidders who believe it easier to fool the government than the purchasing departments of private firms and who are happy to promise more than they can deliver, whether because overeager and undercompetent or disingenuous and deceitful. Ultimately, if indirectly, private companies rely on profitability to deter or detect bribes, kickbacks, and other forms of collusion between their purchasing departments and suppliers. Lacking such measures, government instead has barred its employees from exercising discretion in culling unrealistic bids and incompetent bidders, instead putting in place legalistic hurdles to sort capable suppliers from those who are simply after the public's money. That is the reason for seeming absurdities such as the 18-page milspec fruitcake recipe, MIL-F-1499F.[45] More than one vendor has probably tried to sell the government inedible fruitcake.

Corporate purchasing departments routinely weigh suppliers' past performance. Low bids do not automatically win, the more so given the recent popularity of "relational contracting"—long-term ties with favored suppliers who can credibly promise high levels of quality and dependable just-in-time delivery. By contrast, federal contract law, based on an idealized model of "full and open competition," left officials until quite recently with little latitude in considering past performance. A contract officer might have no choice but to set aside the satisfactory work of an incumbent vendor and contract with a low bidder who performed poorly in the past. The following excerpt from a confidential interview with "a top government marketing manager for one major computer vendor" illustrates the sort of outcome that sometimes resulted:

> Our attitude is: "Bid what they ask for, not what they want." We look at a government specification and see it has loopholes and errors. We don't tell the government that the specification won't do the job. You win the contract, and then you go back to them afterwards, and say, "By the way, the thing you specified won't work. We need a change order."[46]

Firms that dealt this way with private-sector customers would not keep their business long. Similarly, a private-sector manager who approved the award of "a [computer software] program requiring more than 2 million lines of real-time embedded code . . . to a contractor that had almost no meaningful software development experience" would soon be searching for another job.[47]

Congress rewrote the law in the 1994 Federal Acquisition Streamlining Act to permit government procurement officers to consider past performance. While a useful step, this has more relevance to selecting bidders for post office construction or computerized payroll systems than weapons, for which DOD is the sole customer, there are rarely any products on offer from which it can choose, and no more than a handful of firms—sometimes only one—can satisfy service requirements. Under those circumstances, the laws and regulations governing acquisition serve a clear purpose in deterring conduct that market competition weeds out, if imperfectly.

The fundamental reason that "waste, fraud, and abuse" get so much attention is perhaps the opacity of DOD's major programs. Outside observers who grasp neither the technological intricacies of these programs nor their military rationales may perceive corruption where it would be more accurate to see ill-posed, ambiguous, or incoherent mission statements, overly ambitious or contradictory technical requirements and habitual political and bureaucratic maneuver, some of it quite aimless. Contractors do hire retired military officers and retain former members of Congress as lobbyists. Their executives sometimes circulate through high Pentagon positions. Yet contractor influence is easily exaggerated. So many parties are involved in major acquisition programs, with so many cross-cutting interests—not a few of them internal to the services and thus doubly difficult for outsiders to disentangle—that nothing that might be called a military-industrial complex can usefully be said to exist. Congress, the White House, and the armed forces may agree or disagree on the weapons systems the United States should build, but not because some oily arms merchant has misled or corrupted them.

CHAPTER 5

ORGANIZING FOR DEFENSE

Until the latest of our world conflicts, the United States had no armaments industry.
American makers of plowshares could, with time and as required, make swords as well.
But now we can no longer risk emergency improvisation of national defense; we have
been compelled to create a permanent armaments industry of vast proportions.

—Dwight D. Eisenhower, "Farewell Address"
January 18, 1961[1]

The modern defense industry emerged in the 1950s as Washington sold off plant and equipment built up during World War II. By the end of that decade, military production had been largely privatized, with nuclear warheads the outstanding exception. A more diverse set of organizations replaced the "military system of innovation" (to the extent such a label can be justified) earlier centered on the arsenals and supply bureaus of the armed forces and supplemented by inventors and firms that sold specialized equipment to the military. Rapidly increasing technological complexity, exemplified by supersonic aircraft, guided missiles, and computerized command and control networks, sparked widespread adoption in government and industry of new technical and managerial methods—"systems thinking."

The industry that emerged during the Cold War and the relationships that bind it to Washington have few parallels elsewhere in the economy, even though several major defense contractors also operate large commercial businesses. With exceptions for high-volume items such as munitions, defense prime contractors generally build complicated and expensive systems—submarines, satellites, aircraft—in small numbers over lengthy time periods. Rare is the civilian firm that incorporates radically new technologies in a design tailored to the desires of a single purchaser. Nearly 40,000 labor hours go into an AH-1 Cobra helicopter, compared with

25–40 hours for an automobile and a few hundred hours for a heavy truck. Supercomputers and communications satellites for commercial customers are both technology-intensive and labor-intensive, but nongovernment purchasers are far more risk-averse than the Department of Defense (DOD) and much more likely to insist on conservative design choices that promise (if not necessarily deliver) predictable costs and performance. Nothing in the civilian economy resembles the lengthy processes of contract R&D and exploration of conceptual alternatives with which major weapons programs begin.

During two world wars, Washington cobbled together temporary military systems of innovation. The second was far more elaborate than the first, but both were improvisations; there was little more foundation on which to build in 1940 than there had been in 1915, given that hardly any preparations had been made. The system subsequently built in the 1950s, again under the spur of wartime emergency (Korea), solidified into permanence by the end of that decade and has changed relatively little since, despite enormously increased spending.

If the U.S. system of innovation as a whole is viewed in terms of two loosely coupled subsystems, one for commercial innovation and the other for military, the relationships between the two have passed through three stages since the early twentieth century. Until World War II, the armed forces paid little attention to technology and often resisted innovation. Over this same period, organized R&D spread in the civilian economy and innovation became a routine element in the business strategy of large manufacturing firms (chapter 3). The military discovered the value of R&D during the war and, in effect, began learning to innovate systematically in the 1950s. By the end of the 1960s, as we will see in chapter 10, military and commercial technologies diverged, spin-off declined (e.g., from military jet aircraft to commercial airliners), and defense again grew isolated from the rest of the economy. Since the 1970s, the military and commercial innovation systems have drawn well apart.

Privatizing Production

Defense production migrated almost entirely to the private sector over the 1950s. During World War II, government paid for hundreds of new chemical and munitions plants, aircraft factories, and shipyards, and for the retooling of many others. Fearing postwar overcapacity and perhaps a new depression, industry was wary of investing; the Treasury financed about two-thirds of wartime capacity expansion.[2] Government agencies ran some of the new facilities; others were leased to private firms. After the war, government sold much of the new plant and equipment, either for conversion to civilian production or for continued use by defense contractors.

In some cases, wartime spending resulted in entire new industries. With shipments of natural rubber from Asia cut off, Washington paid for 29 synthetic rubber plants operated by 20 different firms. From a near-zero base three years earlier, output reached 1.4 billion pounds in 1945.[3] At war's end, the government owned 95-plus percent of the nation's synthetic rubber capacity, 90 percent of capacity for magnesium, and over 70 percent for aircraft and aircraft engines.[4] To keep up with expanding aircraft production, Washington contracted with Alcoa, the monopoly supplier of aluminum and its alloys thanks to its dominating patent position, to build and operate 20 new plants. During the postwar sell-off, the War Assets Administration permitted Alcoa to buy only one of these plants. The others went to two new entrants, Kaiser and Reynolds. Meanwhile the Justice Department, which had long been trying to break up Alcoa, forced the company to license its patents, which covered both the extraction and processing of primary aluminum and the compositions of some 90 percent of the alloys in widespread use, to its freshly minted rivals at nominal fees.[5] The result was not only an end to Alcoa's monopoly but widespread dissemination of the company's accumulated technical knowledge.

By the end of the 1950s, the public-sector share of total defense output, excluding nuclear weapons, had fallen to perhaps 10 percent on the way to 1 percent at the end of the 1980s.[6] Since then, private firms have also delivered a growing volume of services formerly provided by military personnel— running cafeterias, training foreign militaries, maintaining high-technology weapons systems even in combat zones, standing guard in Iraq. In the 1930s, private arms manufacturers had been suspect, labeled "merchants of death." During World War II, big business, which had taken a good deal of blame for the depression, rebuilt its reputation on wartime production achievements. During the Cold War defense firms became, for most Americans, respectable participants in the "free enterprise" system of lightly regulated capitalism, while government arsenals and shipyards were tagged as harbingers of socialism.[7] With minor exceptions, the United States has kept only nuclear warheads, with their terrifying destructive power, under close control inside government facilities. And even those are managed by contractors.

Although military arsenals still produce a few items, such as artillery tubes, nearly all procurement funds now flow to private firms. R&D dollars pass through more variegated channels. As RDT&E (research, development, test, and evaluation) spending rose following the Korean War, the services strengthened their internal research capabilities, expanded their ties with university research groups, and began to support a wider range of speculative projects conducted by defense contractors that earlier had focused on engineering to the near exclusion of research. Military laboratories worked with industry, sometimes making direct contributions to contractor-developed systems such as towed radar decoys, and with academics who specialized in critical disciplines. University-based materials scientists, for instance, might conduct tightly focused studies of ablative materials for missile nose cones under contract from the Air Force Office of Scientific Research (AFOSR).

In recent years, defense firms have received about four-fifths of RDT&E funds. Most of this goes for the design and development of new weapons systems. Industry's share of research is considerably lower, about half of S&T spending (i.e., budget categories 6.1–6.3, basic research, applied research, and advanced technology development).[8] One-third of S&T funds go to the military's own laboratories, which number around 80, down somewhat since the end of the Cold War, and the rest to universities and other nonprofits. Universities get more than half of 6.1 basic research dollars, but this category makes up only about 2 percent of RDT&E.

Table 3 lists all firms with DOD contract awards that came to $5 billion or more in fiscal 2006. The group includes companies that specialize in defense (Lockheed Martin) and others for which military sales make up a smaller share of revenues (Boeing, especially in the years prior to its purchase of Rockwell's military divisions in 1996 and of McDonnell Douglas in 1997). Table 4 lists DOD's largest unclassified programs for fiscal 2007.[9] As table 5 indicates, aircraft continue to take the largest share of procurement dollars (on-board electronics now accounts for much of their cost).

Table 3   Largest Defense Prime Contractors

	New Contract Awards, Fiscal 2006 (Billions of Dollars)
Lockheed Martin	$ 26.6
Boeing	20.3
Northrop Grumman	16.6
General Dynamics	10.5
Raytheon	10.1
Halliburton	6.1
L-3 Communications	5.2

Source: "100 Companies Receiving the Largest Dollar Volume of Prime Contract Awards—Fiscal Year 2006," U.S. Department of Defense, Statistical Information Analysis Division, undated.

Table 4   Largest Defense Acquisition Programs

Program	Service	Description	Fiscal 2007 Budget (Billions of Dollars)
Ballistic Missile Defense	DOD-wide	Multilayer "system of systems"	$ 9.4
F-35 Joint Strike Fighter	Air Force/Navy	Multirole fighter	5.0
C-17	Air Force	Cargo plane	4.7
F-22	Air Force	Air superiority fighter	4.0
Future Combat System	Army	Networked multimission "system of systems" including air and ground combat vehicles	3.4
DDG 1000	Navy	Destroyer	3.4
F/A-18E/F	Navy	Tactical fighter/attack plane	3.0
Virginia-class submarine	Navy	Attack submarine	2.8
V-22 Osprey	Joint	Tilt-rotor aircraft	2.1

Note: Includes all unclassified programs with 2007 budgets in excess of $2 billion.

Source: "Program Acquisition Costs by Weapon System," Department of Defense Budget for Fiscal Year 2008, February 2007, pp. i–iii.

Roughly half by value of major procurement contracts goes out again in the form of subcontracts (over half for ships, less for tanks and other ground vehicles, with aircraft commonly in the 40–50 percent range), many of these to firms for which military sales are a minor share of revenues. Because of this, defense is perhaps best pictured as a virtual industry. Following a string of post–Cold War mergers, only a handful of prime contractors remain—large companies able to manage the design, development, and production of complex systems more or less in their entirety. These firms orchestrate the work of a large and shifting array of subcontractors and suppliers. Whether or not the latter think of themselves as belonging to the defense industry, they constitute part of the defense industrial base.

With electronics, including software, accounting for something like half of acquisition spending, the merger wave of the 1990s brought further melding of aerospace and electronics capabilities. When Lockheed Martin bought Loral in 1996, for instance, it acquired a company that itself had purchased more than half-a-dozen defense electronics firms earlier in the

Table 5   Procurement by Type of System

	Fiscal 2007 Budget (Billions of Dollars)			
	Army	Navy	Air Force	Total
Aircraft	$ 4.9	$ 10.8	$ 13.9	$ 29.7
Ships	–	10.5	–	10.5
Missiles/Weapons	1.3	2.7	4.0	8.0
Ammunition	2.0	0.9	1.1	3.9
Tanks and Other Vehicles	5.3	–	–	5.3
Other	15.2	11.1	16.7	43.0
Service Totals	*28.7*	*36.0*	*35.7*	*100.4*
Defense Department Total				*$ 103.8*

Note: Totals may not add because of rounding. Navy includes Marine Corps. Aircraft includes engines, electronic systems, and other as-delivered equipment. Defense Department total includes $3.4 billion for agency-wide programs.

Source: "Procurement Programs (P-1): Department of Defense Budget, Fiscal Year 2008," Office of the Under Secretary of Defense (Comptroller), February 2007, p. II.

decade. As the bigger contractors consolidated through mergers, the subtiers contracted. With procurement spending falling in the 1990s, some suppliers stopped pursuing defense orders, preferring to focus on commercial lines of business. Those for which defense sales comprised a small share of revenues were particularly likely to exit. Some of these firms had long complained that burdensome DOD contracting and administrative requirements, lack of incentives for productivity-enhancing investments, and thin profit margins made doing business with government more trouble than it was worth.[10] At the same time, large prime contractors, much like their counterparts in the civilian economy, were streamlining their supplier networks in search of efficiency and "leanness."

Pentagon policies and practices insulate prime contractors from many technical risks and almost entirely from the market risks that commercial firms confront on a daily basis—recessions and price wars, import competition and the vagaries of consumer taste. With DOD watching somewhat fitfully over the industrial base to preserve R&D and production capabilities that might be needed in the future, Washington rescued failing contractors more than once during the Cold War.[11] From time to time, government has spread work among firms and absorbed the cost increases that resulted when gains from competition failed to balance losses of scale and learning economies.[12] In 1993, Congress and the administration agreed to purchase a third Seawolf submarine ostensibly to keep each of the nation's two nuclear-capable shipyards busy. Three years later Congress again put forward industrial base concerns when it directed the Navy to buy Virginia-class submarines from both Electric Boat (a division of General Dynamics) and Newport News Shipbuilding (owned by Northrop Grumman), even though Newport News would have continued to build nuclear-powered aircraft carriers.[13]

New Institutions for R&D and Systems Development

Some of today's defense contractors descend from aircraft firms that won military design competitions in the 1930s and expanded greatly during the war, such as Boeing. Others were formed during the Cold War to support DOD's need for "systems" expertise—not just computer systems, but satellites, missiles, and aircraft of rapidly growing complexity. Tight coupling between a missile's structure and its propulsion system, for example, means the two subsystems must be designed as one. With most of the missile's initial weight consisting of propellant, dynamic response changes dramatically during flight as fuel burns off, possibly leading to vibrations that may threaten to tear the structure apart. Defense firms had to solve new technical problems and manage new sets of technological tradeoffs. The services had to learn to oversee programs in which the contributions of multiple contractors were closely coordinated so the resulting system would function properly. Technological change spurred organizational and sometimes institutional change.

World War II had demonstrated, not least through the horrific effects of the bombs dropped on Hiroshima and Nagasaki, that the risks of innovation would in the future be outweighed by the risks of failure to innovate. It was plain to aircraft firms and others that wished to continue doing business with the Pentagon that they would have to improve their technical capabilities. To remain viable as prime contractors, aircraft manufacturers diversified into missiles and electronics, turning themselves into aerospace firms. They opened research laboratories to complement their design offices, hired newly graduated engineers and scientists trained in gas dynamics, control theory, and solid-state physics. Some built R&D campuses at sites remote from their production facilities and engineering offices. Others opened "skunk works" for secret projects and experimental development.

Although the armed forces exited World War II with greatly increased technical capacity of their own, there was no way they could steer the development of digital computers and guided missiles as their supply bureaus had dictated the design of howitzers and steam turbines. Vannevar Bush's

Office of Scientific Research and Development (OSRD), staffed by some of the nation's best-known scientists and scientific administrators, was shutting down. The services would have to find their own ways of monitoring fast-moving technical disciplines and managing interactions with university groups and private-sector contractors that were developing depth and breadth in the often arcane fields of engineering and science on which national security now depended. The early postwar years threw up a variety of possible models. By the late 1950s, patterns in military support for technology and science had crystallized in much the form they still exhibit today.

In 1946 Congress passed legislation establishing the Atomic Energy Commission (AEC) and the Office of Naval Research (ONR). When the first glimmerings of fission as a possible power source reached Washington in 1939, a few far-sighted officers had urged the Navy to begin work on nuclear propulsion. During the war, the Manhattan Project claimed all available resources. Looking ahead, a small group of relatively junior naval officers began to lay groundwork for what became ONR. Their original intent was to centralize postwar R&D on shipboard power reactors.[14] Almost as soon as ONR came into existence, it lost that responsibility to Hyman G. Rickover, then a captain based in the vastly more powerful Bureau of Ships. ONR managers had to find a new mission. This they defined as support of basic research in almost any field that might be of future interest to the Navy. In the process, the fledgling agency created what became the prototype for federal support of academic research, adopted in similar form by nondefense agencies such as the National Science Foundation (NSF) as well as the Army and Air Force when they set up their own research arms in emulation of ONR. Navy administrators based ONR's operating practices on those of OSRD, which itself had drawn on contracting procedures governing MIT's prewar work for industrial clients and on the practices of the philanthropic foundations that provided most of the external support available for academic research before the war.[15]

By 1947, ONR funds were supporting 2,500 graduate students in 220 university programs.[16] This marked a sharp break with the past. Before the war the military supported very little research, and certainly not in universities, while during the war OSRD, managed and staffed entirely by civilians, paid for essentially all government-funded work conducted by university-based organizations. Now the Navy was paying directly for project-level work in universities, for its own reasons. ONR remained the largest single source of federal dollars for basic research in the physical sciences and engineering until 1965, when finally overtaken by NSF, slower-starting but faster-growing. The Army and Air Force followed the

Navy into support for research (through AFOSR and the Army Research Office). Not until 1973, with the Vietnam War and Mansfield amendment diverting dollars from military R&D, did NSF have more money for basic research than DOD in total.[17]

ONR and its counterparts searched out specialists, many but not all in universities, who could help with narrowly defined problems—algorithms for predicting the propagation of underwater sound, methods for purifying semiconductor materials, mathematical models for heat transfer in jet engine components. Universities had a comparative advantage in tightly focused work that fed the knowledge base for system design and development. Professors and administrators were hungry for money, overhead rates low, graduate students almost absurdly cheap, and both faculty and students could be viewed as recruits in a reserve army available should the Pentagon need their expertise in some future emergency.

Those who oversaw research in agencies such as ONR and AFOSR, whether civilian employees or younger officers (who were more likely than their seniors to be technically trained), usually took for granted that even the most esoteric projects might on occasion lead to military advantage and were in any event necessary insurance against technological surprise by the Soviet Union's own highly capable engineers and scientists. The generals and admirals at the top were not always persuaded. During ONR's early years, high-ranking officers questioned its contributions to practical weapons development (just as their counterparts before the war had doubted the need for the Naval Research Laboratory).[18] On the other hand, Henry H. Arnold, Commanding General of the Army Air Forces during World War II, had often registered his enthusiasm, announcing in 1944: "For twenty years the Air Force was built around pilots, pilots, and more pilots. The next twenty years is going to be built around scientists."[19] Of course he too wanted weapons, not research. It was just that Arnold was able to see a path from one to the other. More common, perhaps, was bemused tolerance of the sort corporate executives so often expressed toward science and scientists during the first part of the twentieth century.

As in large corporations, military expenditures for research were usually small enough to be disregarded. In most years, there was money enough and a "breakthrough," should one occur, would make everyone look good. Even if skeptics, those in high ranks might find it easier to accommodate modest levels of research than applied work that ventured into novel systems concepts which might threaten established roles, missions, and doctrine. From their perspective, research was low in cost and could be fenced off or dismissed as a laboratory curiosity, whereas systems-oriented R&D and conceptual design might open doors better left closed, as illustrated in the late

1990s by "arsenal ship" proposals. A joint undertaking with the Defense Advanced Research Projects Agency, the Arsenal Ship program lasted only 20 months before canceled for lack of support within the Navy.[20]

If ONR established the template for funding of basic research, the Air Force led the way in marshalling advice from outside scientists. In 1944, Theodore von Kármán, who had become a close advisor to the Army Air Forces, led a sweeping study of the future of aviation. When Arnold created a permanent Scientific Advisory Board, von Kármán became its head.[21]

Unlike ONR, Rickover, who became an admiral in 1953 while at the same time serving on the AEC staff, was no model. Irascible, indomitable, and inimitable, he often seemed to act as his own system integrator. Keenly aware from earlier shipboard experience of the many practical issues that would have to be addressed if sailors rather than physicists were to operate nuclear reactors routinely, safely, and with high reliability, Rickover and his hand-picked subordinates pushed research-oriented AEC personnel and laboratories to move beyond science to the practical problems of reactor design. He instigated competition between Westinghouse's Bettis Laboratory and General Electric's Knolls Atomic Power Laboratory and insisted on extensive laboratory testing of prototypes before proceeding to sea trials.[22]

More so perhaps than any other military innovation, even the atomic bomb, the nuclear-powered submarine showed that the time had passed when the United States lagged other nations in the technologies of warfare. Shortly after commissioning in 1955, Nautilus, the Navy's and the world's first nuclear-powered submarine, averaged more than 20 knots on an underwater voyage of more than 1400 miles. This was a speed diesel-electric submarines could barely attain, much less sustain for more than a few minutes.

Rickover's naval reactors group exercised tight control over reactor design and by extension many other features of U.S. submarines. Their record of technical and managerial accomplishment was nearly unblemished. It had to be. Anything less and Rickover would have been forced out long before he finally retired in 1982 at age 82. His methods were deeply resented within the Navy, tolerated only because of the backing he so assiduously cultivated in Congress and among the public. Rickover was one of a kind. Other parts of DOD had to find their own methods of overseeing systems development and integration.

For the Air Force, which struggled during the 1950s to manage work on linked arrays of early-warning radars, computers, and command centers to detect a possible Soviet air attack and on jet-propelled aircraft and air-to-air missiles to deter such an attack or defeat it should deterrence fail, doing so well outside U.S. borders if at all possible, systems integration posed greater difficulties than for the Army or Navy. Arnold might have been a vocal supporter of technology and science, but that had never translated into much in the way of internal capability. In the mid-1930s Air Corps, "[the] project office [serving] as a coordinating point for engineering, procurement, and field operating problems . . . for all bombers . . . consisted of one officer, one civilian engineer, and one stenographer."[23] The officers placed in charge "were often lacking in the personal capacity and preliminary [technical] training desirable," which was one reason "[v]ery few really first-rate engineers remained with the Army very long"[24] Only after the Air Force Scientific Advisory Board issued its 1949 "Ridenour report" on the organization and management of R&D did the newly independent service begin a serious effort to create a technically trained group of officers. Not until 1959 did the engineering-oriented Air Force Academy graduate its first class.

As we have already seen, the Air Corps from the beginning relied heavily on private firms. Failed efforts to build airframes and engines at McCook Field (later renamed Wright Field) at the beginning of the 1920s reinforced the earlier lesson of World War I, that design and production should be left to industrial firms. Still, the technological problems of the postwar years were of a new order and industry too had to find new ways of approaching them. As "steady flow" devices built around continuously rotating compressors and turbines, jet engines had little in common with piston engines, with their reciprocating motions and pulsating or unsteady gas flows. On the one hand, very rapid changes in the conditions inside a piston engine as gases flowed in, fuel burned, and the products of combustion flowed out had made theoretical analysis almost impossible and development almost entirely a matter of trial-and-error. The analogous processes in jet engines were continuous, took place nearer to constant conditions, and therefore could be analyzed with the tools of engineering science. On the other hand, a vast experience base existed for piston engines and the aircraft they powered, hardly any of which was applicable to jets, the more so as they began to approach and then penetrate the "sound barrier." At the speeds now possible, aerodynamic phenomena became highly nonlinear and other technical questions on the standard list, such as flight stability and structural integrity, often in the past handled by rules-of-thumb, needed firmer analytical foundations.

Comparing the supersonic F-4 to the World War II Mustang, fighter ace Robin Olds said, "There was an elaborate strap-in procedure. The F-4 seat alone was more complex than the whole P-51!"[25] Olds exaggerated to make a point. Still, it was no trivial task to integrate an ejection seat with the rest of the plane. Air flows had to be mapped to minimize the risk that an ejecting pilot would be hurled against the tail. Designers had to make sure missiles launched from beneath the wings would not tumble out of control, if only so that the plane itself would not run into them.

Without too much oversimplification, older generations of warplanes could be viewed as airframes designed by one firm mated to engines from another, with guns, bomb racks, instruments, and radios from still others tacked on, most of these components independently procured and delivered for assembly as "government furnished equipment." The new jets, packed with electronics and armed with missiles as well as guns (or in place of them), had to be designed from the beginning as assemblages of interacting and interdependent subsystems. Engineering for a piston-engined fighter such as the Mustang had required something like 15,000–20,000 hours; a dozen years later, jet fighters were consuming 1.5 *million* engineering hours.[26] R&D had to be coordinated and hardware made available on a common master schedule, initially for bench and flight tests and then for production. With many contractors designing parts intended to function together "seamlessly," the Air Force groped to find workable methods of overseeing the entire undertaking, whether doing so itself or by delegating responsibility to a trusted contractor that could be relied on look out for the service's interests above its own.

In the early 1950s, the Air Force was unsure of how best to manage programs in which several contractors untook related tasks. The Atlas missile program, headed by Bernard Schriever—an engineer who entered the Army as a reserve officer in the 1930s, attracted Arnold's attention, and became his protégé—did much to embed systems thinking in the postwar Air Force.[27] Believing the old-line aircraft companies lacked depth and breadth in the technical disciplines needed to develop long-range ballistic missiles such as Atlas, Schriever and the Air Force turned to Ramo-Wooldridge (later Thompson Ramo Wooldridge, TRW), recently started by two Caltech graduates who had worked at Hughes Aircraft. Ramo-Wooldridge grew swiftly—within half-a-dozen years the firm employed over 5,000 people—and became the prototype of the new systems integration firms. Ramo-Wooldridge served, not as prime contractor with overall responsibility for design, development, and production—the old model—but as the agent of the Air Force, overseeing and coordinating the work of other firms. Those being coordinating were not very happy with the arrangement, since they now found themselves reporting, in effect, to Ramo-Wooldridge—a proximate competitor—rather than an Air Force program office. There were also misgivings elsewhere in government. The Air Force responded by barring Ramo-Wooldridge from competing for procurement contracts and, when the Aerospace Corporation was set up a few years later to undertake similar tasks, made sure it was organized on a not-for-profit basis as a federally funded research and development center (FFRDC).

Systems-oriented technical and management methods spread rapidly. In 1959, Schriever became head of Air Force Systems Command, a position he held until retiring in 1966 (as a four-star general). Systems Command had earlier been the Air Research and Development Command; the name change itself constituted a high-level endorsement of systems thinking for everyone in the Air Force to see. Project- and program-level approaches such as those developed for Atlas and the Navy's Polaris program soon met and meshed (or clashed) with top-down methods imposed by the Office of the Secretary of Defense (OSD) under McNamara.[28] The relatively formalized methods of systems management coexisted with, but did not displace, less structured approaches. Bootleg projects conducted out of sight of nominally responsible authorities did not disappear.[29] Nor did the variant illustrated by the Army's development of the attack helicopter (chapter 2). While the officers who instigated and guided this work neither hid what they were doing nor trumpeted it, they had no wish to be exposed to the flow charts and decision points associated with systems management, if only because that might have drawn the attention of possible opponents.[30]

Upheaval in Washington following the 1957 Soviet Sputnik flights, which sparked outraged demands for closer coordination and better management of the tangle of competing rocket, missile, space, and missile-defense programs run by the armed forces and the Central Intelligence Agency, led to the establishment of the National Aeronautics and Space Administration (NASA) and the (Defense) Advanced Research Projects Agency (ARPA or DARPA), largely completing the government's postwar institutional structure for technology and science.[31] Its first director observed that "NASA was born out of a state of hysteria."[32] Cautious and conservative management, more concerned to fend off political opposition than to spark innovation, had left the National Advisory Committee for Aeronautics (NACA) unprepared for the jet age, much less the space age. During the interwar years, smaller aircraft firms with limited technical capabilities charged that NACA R&D amounted to an unfair subsidy benefiting their larger and more technically able rivals, which could more easily absorb and exploit the agency's research results. With complaints reaching Congress, NACA responded by restricting the largest part of its technical agenda to generic design information that almost any firm could understand and apply, such as compilations of the lift and drag characteristics of airfoil cross-sections and the strength properties of standard-composition aluminum alloys. NACA avoided theoretically oriented work of the sort that had become prominent in Europe, especially in Germany, and was beginning to find a place in American universities (e.g., at Caltech, where von Kármán transplanted the German approach). NACA from its founding had also neglected propulsion, one reason why defense officials during World War II, unwilling to rely on the organization for a critical new technology, cut the agency out of jet engine work.[33] Instead, the Army turned to firms including General Electric, with its well-regarded research capabilities and experience with steam turbines and turbo-superchargers for piston engines. NACA was relegated during the war mostly to testing planes and power-plants for the military. Afterward, even as technical data brought back from Germany revealed how badly the United States had lagged in aerodynam-ics, NACA's self-selected field of emphasis, the organization relapsed into its prewar sluggishness.[34] Amidst the uproar caused by Sputnik, there was no confidence in Washington that NACA could step in to lead or coordi-nate a response.

With the establishment of NASA in 1958, the new agency absorbed NACA's laboratories and most of its employees. NASA functioned quite differently. Whereas NACA had been strictly an R&D organization, one that conducted almost all of its work internally, Congress assigned NASA responsibilities that ranged far beyond R&D and from the beginning the new agency relied heavily on contractors. NASA did resemble NACA in being an integral part of the "national security state." Policymakers trans-ferred the Army's ballistic missile programs, which had been developing powerful rockets that promised a quick route into space, to NASA, along with a number of Air Force and Navy rocket and satellite projects. NASA collaborated with defense agencies on the technical problems of supersonic flight and fly-by-wire controls, critical for stealth aircraft such as the F-117 and B-2.[35] Not only does the space shuttle place satellites in orbit for DOD and intelligence agencies, one-quarter of the $1.7 billion expended on the National Aerospace Plane (NASP) came from NASA (DOD provided the rest). Advocates hoped that NASP would demonstrate the ability to fly payloads for the Strategic Defense Initiative directly into orbit (i.e., with-out booster rockets).[36] NASA has been as close a partner of the military as NACA ever had been.

Before Congress passed NASA's enabling legislation, DOD had put in place its own response to Sputnik. Secretary of Defense Neil McElroy pushed DARPA into being over the opposition of the services, which had no wish to cede control over any portion of military R&D, and certainly not to the civilians slated to run the new agency.[37] McElroy gave DARPA control over military missile and space work not intended for NASA. With OSD seeking to tamp down the interservice rivalries that had plagued missile and antimissile work the services were able to claw back some of these projects. Unable to hold onto their portfolio, DARPA managers, like their counterparts at ONR a dozen years earlier, had to find a new mission. This they did by becoming the home within DOD for long-range applied research and farsighted technology development.

DARPA has prospered by filling "gaps" between relatively fundamental research and systems development. Such gaps tend to open when none of the services steps forward as primary customer for a new technology or when the technology is applicable to all, as with digital electronics, thus calling for a coordinated effort of the sort that has always been so difficult for the armed forces. In recent years known to the public especially for nurturing the Internet's predecessor, the wide-area computer network known as ARPANet, the agency undertakes projects that are small in scale by DOD standards, with applications that may be well in the future. As a result, DARPA has sometimes had to walk a fine line to justify the military relevance of its work. The agency has been through a number of cycles in which tolerance for visionary programs ended with a turn toward near-term applications and prototype demonstrations. At other times, depending on the views of DARPA's own managers, elsewhere in DOD, or in Congress, the agency has felt freer to invest in technologies for which future applications were undetermined or indistinct.

More so than DOD's research agencies, such as ONR and AFOSR, DARPA operates as OSRD did in World War II. Rather than long-term research intended to generate new knowledge in support of military needs, much of this research in established disciplinary fields of engineering and science, DARPA has focused on weapons themselves. Vannevar Bush and OSRD pushed constantly to get new weapons and weapons-system components out of the laboratory and into the field because there was a world war underway. DARPA did so because there was a Cold War ongoing. This kind of work is multidisciplinary, well removed from the agendas of scientific and technical communities populated by research-oriented academic scientists and engineers. Yet much of it remains too far from realization to attract sponsorship by the systems-oriented branches of the services.[38] Like OSRD, DARPA has relied heavily on the judgment of a highly qualified technical personnel. DARPA's staff is small and turns over rapidly as outside scientists and engineers, many of them well-known in their fields, cycle through on short-term appointments. Perhaps the biggest difference is that the bulk of DARPA contracts go to private firms, while OSRD favored R&D organizations based in universities, in part because there was hardly any research capability in the defense industry at the time of World War II.

Alongside the military's own R&D organizations, about three dozen federally funded research and development centers, as well as other not-for-profit organizations such as the Johns Hopkins Applied Physics Laboratory (APL), conduct work for DOD and the rest of government. MITRE Corporation specializes in large-scale computer systems. Since the 1940s, APL, once but no longer an FFRDC, has worked on missiles and electronic systems for the Navy.[39] FFRDCs run the nation's nuclear weapons laboratories for the Energy Department. Several, such as RAND Corporation, serve as "think tanks" for the military, bringing perspectives free of the influence of service culture and the profit-seeking motives of defense firms.

Essentially all the new organizations and institutions sketched above were in place by the end of the 1950s. At that time, the nation's overall system of innovation could be represented as two linked subsystems, one for military innovation and the other for commercial.

Two Systems of Innovation

That structure was fundamentally new. During both world wars, policy-makers assembled military systems of innovation on an ad hoc basis.[40] Military R&D barely existed before 1940. The defense industry consisted of a few small firms that supplied instruments such as bombsights and gun directors to the Army and Navy, somewhat larger firms that built aircraft in batches of a few dozen, or produced other matériel to the specifications of military supply bureaus. Innovation was not a policy objective. When it occurred, it was likely to be the result of individual rather than organizational initiative. By contrast, industrial R&D in 1940 had become an integral part of the business strategies of the large manufacturing firms that dominated the nation's economy, a result of conscious managerial decisions going back half a century.

The commercial system of innovation of the 1950s, in other words, had grown up more or less organically, with contributions from many sources over many years. The military subsystem had been created in Washington over a period of a dozen years, 1946–1958. It was based on the experiences of World War II, but those experiences had not been fully digested when the Korean War initiated another burst of improvisation. Since then, the commercial system has continued to evolve at a rapid pace as the civilian economy shifted away from the manufacturing-centered structure of the first part of the twentieth century, in which R&D initially came to prominence, and toward the service-dominated structure that has drawn so much attention since the 1980s, in which formal R&D is less visible.

In some respects, the post–Korean War improvisations were a resounding success. DOD constructed institutions that for a time were vital elements in a national innovation system that helped make the United States preeminent in the global economy during the Cold War and remains the envy of the rest of the world today. In search of technological advantages vis-à-vis the Soviet Union, the Pentagon spent huge sums on R&D and procurement, supporting a technology base from which commercial firms, always hunters and gatherers of technology, could feed. During the 1950s, defense put a deep and lasting imprint on the U.S. national innovation system. Later the world of defense grew isolated, a result of both technological and economic change. By the 1980s, the military had lost much of its ability to exploit the national system of innovation. As applications of technologies such as digital electronics exploded in the civilian economy, DOD increasingly was left to its own devices, ignored by innovators outside the specialized defense industry. Long before the business press announced a "new economy" in the 1990s, the Pentagon had come to seem mired in the old economy, sclerotic by comparison with the entrepreneurial vigor for which Silicon Valley was so widely celebrated. The irony, of course, is that Silicon Valley had been built to considerable extent on the defense spending of the 1950s and 1960s.

CHAPTER 6

BUYING MILITARY SYSTEMS: WHAT THE SERVICES WANT AND WHAT THEY GET

We invented the term "imprecisely located target." Why would you ever want to send an airplane with a man in it? If you knew the coordinates, you might as well do it with a ballistic missile because it is cheaper and better. There had to be a strategic role for the long-range strategic combat aircraft and that was where the imprecisely located target concept came in.

The result of that study was that the B-1 received revived respectability . . . .

—Ivan A. Getting, interview[1]

Design tradeoffs determine the performance of weapons, just as they do for other technical systems. To prosper, commercial firms must learn to manage these tradeoffs. The Department of Defense (DOD) has no such incentives. It does not and cannot manage tradeoffs well because political-bureaucratic considerations overwhelm those of technology, and often doctrine as well.

Much of this chapter is given over to the institutional setting for weapons development, beginning with competition for budgetary appropriations, which conditions almost everything that happens in defense acquisition. The chapter then turns to a brief discussion of two submarine programs, the Polaris and Virginia, before moving on to air power, a short summary of the development of the assault rifle eventually designated the M16, and the B-1B bomber.

Strategic bombing doctrine as formulated by the Air Corps between the wars assumed that heavy bombers would be able to penetrate enemy-controlled airspace without fighter escorts. There could be no basis in World War I experience for this assumption—aviation technology (along with radio and radar) had advanced too fast and too far for meaningful extrapolation—and World War II would show it to be false. Yet the chosen doctrine of Army aviators was still more deeply flawed, since it was also based on gross overestimates of the damage that strategic bombing could do to the warmaking potential of an enemy state. Further miscalculations followed. In the 1950s, looking ahead to a possible war with the Soviet Union, DOD planners opted to arm jet-propelled fighter-interceptors with guided missiles in place of guns. These planes were intended to streak out from the nation's borders at supersonic speeds, downing Soviet bombers before they could approach U.S. airspace. In Vietnam, American pilots found themselves instead, as in previous conflicts, engaged in dogfights with enemy planes. Their air-to-air missiles, designed to bring down lumbering bombers at long range, proved ineffective in close-quarters combat against rapidly maneuvering enemy fighters.

The B-1 was designed to be anything but an easy target. But the Air Force could not decide what else the B-1 should be. In asking the plane to maneuver at treetop levels and also to fly fast and high, it imposed conflicting requirements that left the B-1B a crippled airplane. While the M16 could hardly differ more as a technological artifact, the two programs had in common the distorting effects of doctrine and mission definition on requirements. As the Army struggled to find a suitable replacement for the World War II infantryman's rifle, it demonstrated that even the simplest-seeming weapon can fall prey to the pathologies of acquisition.

Competition for Dollars

Defense firms have only one customer, the federal government (foreign sales are subject to Washington's approval). But government is not a unitary customer. The services compete for acquisition dollars and they also compete internally—for submarines as opposed to aircraft carriers, for tanks as opposed to helicopters, for fighters as opposed to bombers. These rivalries often have greater effects on technological choice (and force structure) than rivalries among the services (e.g., between Air Force and Navy for aerospace missions that either in principle could perform). The Office of the Secretary of Defense (OSD) maintains shaky civilian control over all this, Congress often becomes involved, the White House is a party to big decisions and sometimes becomes involved when quite prosaic undertakings go off the tracks: President John F. Kennedy intervened personally in the M16 rifle program. In the field, militaries insist on clear-cut, unambiguous chains of command because lives depend on it. No such imperatives exist in acquisition.

Programs compete constantly for dollars. Simply to win initial approvals requires a powerful coalition. One of the services must want the system—and want it more than others vying for a share of that service's acquisition budget. Panels, advisory bodies, and committees must sign off. OSD has to be persuaded, along with the Office of Management and Budget and Congress.

> [T]o get a proposed C^3I [command, control, communications, and intelligence] system through the evaluation phase, it is necessary to orchestrate the inputs of at least 20 different organizations. Simply managing the logistics of such a process requires organization and entrepreneurship.[2]

Requirements are central to the process. Based, at least in principle, on both perceived threats and emerging technical capabilities, statements of mission needs and requirements (which come in several varieties, with names such as "initial capabilities document") provide the summary justification for a program and inform potential contractors of what the initiating service wants. Once the responsible officials have signed off on technical requirements, they become part of the legally binding agreement that follows solicitation of proposals and contractor selection.[3] At this point, the feasibility of meeting these requirements is as likely to have been assumed as demonstrated, given that the analytical studies needed are likely to be so costly and time-consuming that they must await contract awards.

Winning the initial approvals is not enough.

> DOD starts more programs than it can afford and rarely prioritizes them for funding purposes. The result is a competition for funds that creates pressures to produce optimistic cost and schedule estimates and to overpromise capability.[4]

Supporters must go to bat for their program, year after year against all others, to keep the money flowing. The resulting pressures mean that any and all information made available to Congress and the public, and often to OSD as well, is likely to be slanted and shaded. Contractors promise what they cannot deliver and program offices endorse those promises even if they disbelieve them. "According to numerous Air Force and DOD sources, cost assumptions were altered on what one termed a 'lie-until-you-believe' basis."[5] As a result, only those directly involved in a program may know the best and latest cost estimates and schedule projections, which are likely to be hidden from outsiders. DOD's most recent Quadrennial Defense Review notes "deep concern in the Department of Defense's senior leadership and in the Congress about the acquisition processes. This lack of confidence results from an inability to determine accurately the true state of major acquisition programs when measured by cost, schedule and performance."[6]

Presumably DOD's senior leadership knows the reasons but not what to do about it.

Major acquisition programs are overseen by fewer than 1,000 senior managers, many of them military officers on temporary assignment. Officers of relatively high rank (e.g., colonels or Navy captains or above) head most program offices. They oversee staffs that number from a few dozen to a few hundred personnel, both military and civilian. Even a general or admiral heading a program office has little real authority. He or she generally cannot select program staff and may be unable even to decide where to "seat personnel in our office space."[7] Turnover is high. Six generals headed the Army's main battle tank program during the 18 years that elapsed between the initial attempts to define parameters and delivery of the first M1 Abrams. Following legislation in 1990 to create a formal corps of acquisition officers, average turnover has come down somewhat. Even so, the Joint Strike Fighter, a program that began in 1996 and will run at least to 2013, was on its fifth manager by 2005. And while the Army would not send a battalion into the field under an officer who had not proven himself at lower levels, the services exhibit few such compunctions when it comes to acquisition. For an officer detailed as a program manager, there is little to gain and much to lose. The road to the top passes through operational commands. Acquisition is a detour. Why chance derailing a career when one expects to be far away by the time the system finally rolls out and begins its field trials? Managers on short-term assignments have little reason to ask hard questions, demand forthright answers, or tell the truth in public.

Over the years, extensive formal processes, descendants of the Planning, Programming, and Budgeting System (PPBS) instituted by OSD under McNamara, have been put in place in efforts to bring accountability to major programs. PPBS has sometimes been derided as toothless. In his second tour as Secretary of Defense, Donald Rumsfeld called it "an antique" that "works poorly."[8] Within a year he was forced to conclude that "there is no practical alternative to PPBS." Rumsfeld contented himself with changing the name to Planning, Programming, Budgeting, and Execution system (PPBE), to emphasize the need for follow-through once programs had been planned, programmed, and budgeted.

Charles E. Wilson, Secretary of Defense from 1953 to 1957, had warned his successor that "[W]ith a few exceptions in the higher ranks, it was pretty hard to get those fellows [military officers] to think about defense programs in terms of the aggregate national problem, as distinct from the interest of their services."[9] McNamara's objective with PPBS was to ensure that weapons programs reflected what was actually needed, as opposed to what the services wanted. Because "a proper balancing of all the elements . . . can only be achieved at the Department of Defense level," PPBS would become a vehicle for assessing service requests in light of "the defense effort as a whole."[10] PPBS/PPBE remains the primary mechanism for trying to achieve that.

In principle, the Joint Chiefs of Staff (JCS) are responsible for harmonizing the desires of the services and forwarding recommendations to the Secretary of Defense. In practice, the chiefs speak for their services and the JCS aims for unanimity. The usual result is a lowest-common-denominator set of compromises. As explained by General David C. Jones, JCS chair from 1978–1982, "the joint system usually recommends that the secretary [of defense] give the services what they want, but states that they need even more money than they have asked for."[11] Planning may be equally incoherent inside the services. Air Force Major General David McIlvoy recalled, "We produced great, beautiful plans, and the [budget] programmers simply ignored them."[12] That, in short, describes how the United States chooses the weapons it buys. Multibillion dollar decisions are left to the internal politics of the services, struggles carried on behind closed doors where they are largely hidden from civilian officials, including many of those within DOD.

Institutional Rivalries: Submarines and Air Power

Inter- and intraservice rivalries sometimes lead to overlap, duplication, and disarray. Competition can also benefit national security, helping ensure that good ideas get a hearing and bad ideas get exposed. Competition with the Air Force spurred the Navy's Polaris submarine and missile to rapid completion. Among the other lessons of this program, Polaris showed the importance of appropriate requirements. Technical and mission objectives were stated quite generally. The Navy established realistic budgets and schedules and by-and-large kept to them. The Polaris submarine was based on an existing hull class and incorporated proven propulsion equipment. Perhaps of greatest significance, Polaris was the Navy's counter to efforts by the Air Force to monopolize strategic deterrence, which meant the top priority for Navy leaders was to get a functioning system to sea quickly—a motive sufficiently powerful to override the lure of absolute best performance.[13]

The development of the F-117, the first stealthy production airplane, teaches generally similar lessons. The program had a single overriding design objective, to minimize the F-117's radar signature, supported at high levels by OSD officials with the background to understand the relevant technologies and the good sense to avoid interfering once work began. Tight secrecy also appears to have contributed.[14] The fundamental difficulty in replicating this or other examples of effective management is the lack of sound recipes for doing so. Just as in the civilian economy, recipes end up sounding much the same and dismayingly like descriptions of sloppy management: only the outcomes differ, for reasons that may not be evident except in hindsight and even then may be attributed to the skills of particular individuals. Secrecy, notably, too easily becomes a screen for hiding a program's shortcomings, as it did for the Navy's stealthy A-12.

If Polaris was a case in which interservice rivalry proved beneficial, intraservice rivalry exerted opposing effects. Because Polaris took money from the Navy's attack submarines

> [S]enior submariners, with very few exceptions, strongly opposed the program. One had even told junior officers on his staff that their naval careers would be finished if they had anything to do with [Polaris].

The reason?

> [S]ubmarines were meant to sink ships with torpedoes, not to blast land targets with missiles; submarine warfare was a battle of wits against an opponent and not a demonstration of technological sophistication.[15]

Whatever the Navy might have learned from Polaris, finally, did not transfer to the later Trident program, which so overpromised on budget and schedule that delivery of the first hull had to be postponed six times.

Although the close control maintained by Rickover over all aspects of nuclear propulsion contributed to the success of Polaris, within a few years the technological choices he and his staff imposed were causing dissatisfaction among submariners—a case of intraservice conflict that appears to still be playing out, decades later, in the Virginia-class submarine program. The Virginia is to replace the Navy's current mainstay, the Los Angeles class of attack submarines. The design of the Los Angeles was laid down in the late 1960s, with major modifications a decade later. At first, the Seawolf was to be the Los Angeles's replacement: the Navy commissioned the last of 62 Los Angeles hulls in 1996 and the first Seawolf a year later. Designed to hunt Soviet submarines in deep water and beneath polar icecaps, Seawolf was very sophisticated and very expensive. RDT&E exceeded $10 billion and procurement had been put at $1.85 billion each for a run of 30 submarines. With the end of the Cold War and Russian submarines deteriorating as they sat at dock, only three Seawolfs were built. Instead, the Navy has now begun to purchase a planned run of 30 submarines of a newer design, the Virginia. Intended to be less expensive, when planning for the Virginia began in 1991 the target price was to be half that of a Seawolf.[16] The Navy and its contractors have not come close to reaching that goal. Recent estimates put the cost of each Virginia submarine at about $2.7 billion.[17]

The Navy will retire Los Angeles hulls ahead of schedule as newly built Virginia submarines join a shrinking overall fleet. But with Russian sea power having almost literally dissolved, no navy in the world poses a threat even to older Los Angeles submarines. Why then does the United States need a new family of attack submarines? The Navy's answer is for power projection ashore from littoral (shallow inshore) waters, "where the clutter of friendly, enemy, and neutral coastal trade, fishing boats, oil rigs, small islands, dense air traffic, large commercial ships, and an intricate tangle of electronic emissions all create a confusing environment . . . ."[18] Given a future in which no other navy seems likely to mount a threat in the open ocean, it is hardly surprising that the littoral mission has gained a central place in the Navy's post–Cold War vision statements. Still, it is not difficult to find an apparent subtext in the Navy's plans for early replacement of Los Angeles submarines. The Los Angeles reflects Rickover's near-obsessive control over all aspects of propulsion, which spilled over to affect many other design features. From the very beginning, in the 1940s and 1950s, Rickover insisted on conservative design choices to ensure safety and reliability, since sailors would spend months at sea in close proximity to potentially lethal radiation, no quick escape would be possible in the event of an accident, and a nuclear-powered submarine would be close to helpless if the reactor had to be shut down. By the 1970s, many U.S. submariners had come to believe U.S. submarine designs were needlessly conservative, handicapping them vis-à-vis a Soviet submarine force that willingly tolerated greater risks (as attested by much higher accident rates). These sentiments intensified as U.S. intelligence learned more about the USSR's high-performing titanium-hulled Alfa class boats. Although Rickover's "technological style" might have been appropriate in the early years of the nuclear navy, to his critics it had become a burden by the time of the Los Angeles.[19]

Put more crudely, American submariners simply did not much like the Los Angeles. With the Cold War over, and large numbers of post-Rickover Seawolfs impossible to justify, the Navy's fallback became the new Virginia rather than existing and possibly refurbished Los Angeles submarines. It is by no means clear that the Virginia offers sufficient advantages in littoral warfare to justify the costs of an all-new class. Yet according to some in the Navy, not only does the United States need the Virginia, but also needs these submarines in greater numbers than planned: "The only long-term solution [to meeting the 'projected threats of 2015'] is to build the new Virginia class at a rate at least double what currently is authorized."[20]

Within the Army and later the Air Force, advocates of fighters and tactical aviation in general have often been at odds with those who favor steering available funds to long-range bombers. The bomber faction won most of the battles from the early 1920s, when General William "Billy" Mitchell choreographed his famous ship-sinking demonstrations, through World War II. "By 1935 the full-blown theory of high-level, daylight precision bombardment of pinpoint targets was being taught at the [Army Air Corps] Tactical School."[21]

As we have seen, the apostles of air power believed that future wars might be won by strategic bombing alone. They had become "obsessed with the idea that airpower could leap over opposing armies and navies and strike at . . . a hostile nation's economy . . . ."[22] Air Corps leaders ignored or marginalized those within their own ranks, such as then–Captain Claire L. Chennault, a pursuit instructor at the Air Corps Tactical School, who protested that the assumptions underlying strategic bombing were unfounded.[23] Nor was Air Corps doctrine exposed to public scrutiny. Through the 1930s, as we saw in chapter 4, the Army took care to avoid all mention of offensive missions for reasons of domestic politics. Coastal defense became the official rationale for heavy bombers. Planes such as the B-17 would fly out to sea and sink the ships of any enemy presumptuous enough to threaten invasion. This was a façade. Air Corps leaders knew that strategic bombing, and only strategic bombing, could be presented, when the time was right, to higher authorities, the White House and Congress, and to the public at large, as a mission that should properly be entrusted to aviators alone, a mission that justified, indeed required, independent status. That had been their goal since World War I: an Air Force coequal with the senior services.

The effects of strategic bombing were indirect. Since the bombing would have no immediate connection with front-line fighting, those in command of ground troops had no right to a voice in choosing targets. Tactical aviation, on the other hand, required "perfect sympathy between the ground and air commander" to be effective.[24] This meant ground officers had a legitimate claim to at least share in control. Much the same was true for interdiction (e.g., attacks on forces headed for the front) and strikes directed at "assets" in rear areas such as ammunition dumps and fuel storage facilities. In building their case for autonomy, Air Corps leaders downplayed all such missions and drew a sharp line between theater objectives and strategic bombing, which they held to be the most "cost effective" application of air power. The 1943 field manual *Command and Employment of Air Power* stated flatly that "[M]issions against hostile units [near the front lines] are most difficult to control, are most expensive, and are, in general, least effective. [O]nly at critical times are [such] missions profitable."[25] Tactical missions would only dissipate the war-winning potential of air power, scattering planes and pilots (in "penny packets" as it was sometimes put in Britain's Royal Air Force) to meet the demands of hard-pressed troops. If allowed to become the servant of slogging infantry, earthbound armor and artillery, air power would be no more than an extension of existing warfighting practices rather than a pathbreaking innovation in its own right. (At times, aviators rejected the very word "support" as implying subsidiary.) This was premature. At the time, strategic bombing was just a theory. Even today, its effectiveness remains controversial.

Doctrine dictated resource allocations. Massive blows at the enemy's war-making capacity called for a massive fleet of bombers. These planes merited first priority on funds. Believing that they could evade detection or defend themselves if spotted, Air Corps leaders declined to purchase long-range escort fighters to accompany the bombers to distant targets. In 1939 and again in 1941, they went so far as to bar development of range-extending drop tanks. The primary mission of fighter-interceptors was to rise against the enemy's own bombers should they attack, a mission that required no great combat radius. If drop tanks were available, fighters might be diverted into "improper tactical use," meaning ground support—not their assigned role and not to be permitted.[26] And because doctrine proclaimed interceptors to be ineffectual against bombers, the United States permitted itself to fall behind during the 1930s in developing fighters. In other countries too, advocates of strategic bombing argued, against the opposition of general staffs—"dogmatic old men who considered the airplane only a kind of extended artillery"—that air power was "indivisible" and could not reach its potential if tied to ground forces.[27] But only in the United States and Britain did they win out, with the result that, while German military planners had made tactical aviation an integral part of blitzkrieg tactics, the Army Air Forces "entered [World War II] completely unversed in close air support, and lacking worthwhile experience, equipment and doctrine."[28] Britain, close to the Continent and vulnerable to bombing, had at least developed capable fighter-interceptors.[29]

Wartime experience as summarized in chapter 9 would prove the faith of U.S. Army aviators in precision bombing and self-defense by bombers to be ill-founded. When it came time to implement Air Corps doctrine by sending waves of B-17s and B-24s against targets in occupied Europe and

Germany, the planes could not find their targets and could not hit them when they did. Nor could they fend off the Luftwaffe's interceptors: even if well armed and armored and flown in close formation for mutual supporting fire, bombers could not survive in enemy-controlled airspace. Regardless of setbacks, Army aviators remained true believers in the doctrine they held to so firmly.[30] That doctrine then transferred more-or-less intact to the independent post-1947 Air Force. By then, of course, it could be buttressed by appeals to the devastating power of the atomic bomb, which would hardly need to be delivered with pinpoint accuracy to destroy a ball bearing factory or an oil refinery.

After the war, participants at the 1948 Key West meeting on the roles and missions of the services, agreed, in language that might have been pulled from prewar documents, that "strategic air warfare" was "designed to effect, through the systematic application of force to a selected series of vital targets, the progressive destruction and disintegration of the enemy's war-making capacity to a point where he no longer retains the ability or the will to wage war."[31] If World War II had shown that it was difficult even to locate targets, much less hit them, given nuclear warheads "strategic" attack no longer depended on the ability to put a bomb in the proverbial pickle barrel. An entire industrial district might now be obliterated with levels of accuracy the Air Force could actually hope to achieve. As the laboratories of the Atomic Energy Commission (AEC) designed smaller, lighter, and cheaper warheads and produced them in ever-larger numbers, the U.S. nuclear stockpile, which amounted to only 9 bombs in 1946 and 50 in 1948 grew to 1,000 by the close of the Korean War on the way to over 30,000 in the late 1960s. From the 1930s until it became plain that intercontinental ballistic missiles (ICBMs) could deliver nuclear warheads with greater reliability, the strategic bombing mission shaped aircraft design choices in the United States.

Although fighters continued to be tarred as defensive rather than offensive tools, the stigma could be reduced by giving them greater firepower, culminating in nuclear weapons of their own. By 1952 the AEC's nuclear weapons laboratories had developed tactical nuclear bombs that could be carried by the F-84. As the decade continued, the Air Force spent heavily on high-speed fighter-bomber-interceptors of the so-called Century Series, beginning with North American's F-100, the world's first operational supersonic fighter. While the chief task of the Century Series remained defensive, to blunt an incursion by the Red Army into Western Europe or, in the event of a direct attack on the United States, streak out to meet and down Soviet bombers before they could get anywhere near the nation's borders, Century Series planes, including the F-105s that performed so poorly in Vietnam, were also designed to undertake offensive missions. Accompanying the bombers of the Strategic Air Command (SAC), they would surprise Soviet planes on the ground and destroy them before they could take to the air. This was one of the oldest tactics of aviation, proven in World War I, used to devastating effect by Germany in 1941 (when a Luftwaffe force numbering fewer than 1,400 planes had destroyed, in one week, over 4,000 Russian aircraft), and in the later stages of World War II turned to great effect against Hitler's own air force. With one of the smaller, lighter nuclear warheads now available, a single fighter-bomber might destroy an entire air base, a tactic that entire squadrons had often failed to achieve in attacks on the Luftwaffe's airfields. (The cratered runways left by conventional munitions could be repaired quickly.) The F-106, more than twice as expensive as its predecessors, packed with electronics, and designed to accommodate nuclear-tipped air-to-air missiles capable of expunging Soviet planes from the skies, took the Century Series to its 1950s technological extreme. During the 1950s, the primacy accorded nuclear strategy shaped the design of the Air Force's front-line fighters.

In preparing for a war with the Soviet Union that never came, the United States found itself less than ready for the war in Southeast Asia into which it stumbled. None of the Century Series fighters served very well in air-to-air combat, much less ground support (a very different task from attacking fixed targets such as Soviet airfields). In Vietnam, the Air Force lost 383 of its 833 F-105s, as these planes became "dogmeat . . . for [a] little league country."[32] Airpower historian Richard Hallion writes: "[T]he Air Force's . . . conceptualization of the fighter's future role was completely out of sync with . . . every air war since the fighter had first appeared in 1915."[33] His point: the Air Force should have anticipated a continuing need for air superiority fighters, a need that had been amply demonstrated only a dozen years earlier in the Korean War.

As the war in Vietnam continued, the Air Force had little choice but to switch to F-4s originally designed for the Navy. In part because Navy pilots had to fly from the rolling decks of aircraft carriers, the F-4 was more agile than existing Air Force fighters and thus better able to dodge ground fire or enemy MiGs. The F-4 was not only manifestly superior for Air Force missions in Vietnam, but was also in production and available, whereas the Air Force would have had to wait years for a new plane of its own. Even so, OSD had to strongarm the Air Force into buying a Navy plane.[34]

In the end, the Air Force purchased three times as many F-4s as the Navy. Yet the F-4, like several of the Century Series planes, had been designed around air-to-air missiles and did not carry a gun. U.S. missiles, furthermore, had been designed to shoot down Soviet bombers at high altitude and long range, not close-in, fast-turning MiGs. During 1965–1968, American pilots fired 442 missiles in the course of air-to-air engagements, downing 52 enemy planes.[35] Until McDonnell Douglas could begin producing the reengineered F-4E with an internal gun mount, F-4s were retrofitted with an external pod carrying a 20 mm cannon.

The Navy had purchased the F-4 because it needed a capable air-to-air fighter to protect its ships. Almost any mission the Navy might contemplate required shielding its battle groups from air attack. For the Navy, there was no onus on defense and tactical doctrine and training had higher priorities than for the Air Force. In addition, the Navy felt a continuing obligation to provide air support for Marine Corps ground forces, which, because of their amphibious landing mission, were lightly equipped with armor and artillery compared to the Army. While the Corps had its own tactical air units, fully committed then and now to ground support, Marine pilots did not routinely train to fly from carriers until late in World War II. Thus it had been left to the Navy to provide close air support during the island-hopping campaigns in the Pacific. Once airfields had been captured or constructed, moreover, "In technique, the AAF [Army Air Forces] in the Pacific had to learn close air support from the Marines . . . ."[36] After the war, the Navy, with a strong push from Congress, continued to buy planes suited to ground support and put more effort than the Air Force into training pilots for these missions.

During the Korean War, the Navy's new jets had performed poorly compared to those of the Air Force. The reasons had little to do with roles and missions; they were a consequence of the technological limitations of the time. Early jet engines came into their own in the thin air of the upper atmosphere. They could produce little thrust at low altitudes, which made planes sluggish and hard to maneuver during takeoffs and landings. The Air Force could tolerate this since its fighters flew from many of the same bases as bombers, which had always needed long takeoff runs. But first-generation jets were marginal for carrier takeoffs and, especially, landings (catapults boosted takeoff speed; pilots had no such aid in climbing out from a missed landing). That meant the Navy's first-generation fighters had to sacrifice performance at altitude simply to make carrier operations feasible. The Korean War showed how badly the Navy needed fighters that could take better care of themselves once airborne. The second-generation F-4, designed in 1953–1954 with a powerful new engine that had been intended for one of the Air Force's bombers, was the response.

During SAC's heyday in the 1950s, the bomber camp got pretty much what it wanted: B-36s, B-47s, B-52s, B-58s. Technological change then intervened in the form of the ICBM. While the first atomic bombs weighed more than 5 tons, too much to be lifted by the rockets of the time, weight and size came down rapidly. By 1954, the United States had tested nuclear-capable ballistic missiles. In 1957, when the second Soviet Sputnik carried more than 1,000 pounds into orbit, the entire world realized that missiles would be able to deliver nuclear warheads almost anywhere on Earth. Computers, cheap and expendable, flew ICBMs, replacing pilots, navigators, and bombardiers too highly trained at too considerable an expense to be considered expendable by military minds, whether or not those minds cared a whit for the men under their command. The missiles could also be hidden in underground silos or aboard submarines. In contrast, SAC's bombers were potentially vulnerable, especially on the ground, with consequent pressure to "use them or lose them" should hostilities build to the point of war.

In the early 1970s, technological change began to threaten bombers in another way. Precision weapons, by making possible highly accurate delivery of nonnuclear warheads, transformed tactical aircraft into viable strike platforms. Only with luck could the unguided bombs of World War II and Korea hit and damage point targets such as bridges or factories. This meant that huge fleets of heavy bombers, each carrying many warheads, had to be sent to blanket the entire area. Maybe one or a few would hit the intended objective. Smaller planes could not carry even one of the biggest bombs (thus dive bombers could not accomplish much except against troop formations and motor vehicles). Depending on the target, with precision guidance fighter-bombers might not need to.

During World War II, the Army Air Forces had managed to destroy several bridges in Burma with radio-controlled bombs. Two decades of largely incremental advances in sensors and guidance followed. In the 1960s, the Air Force began to support R&D on laser guidance and by the late stages of the Vietnam War laser-directed bombs were destroying targets that had survived hundreds of earlier sorties—871 in the much-advertised case of the Thanh Hoa bridge, finally taken down in 1972. After four decades, the "surgical strike" had been transformed from journalistic trope into military reality. Tactical aircraft could now attack point targets with reasonable prospects of success. With multiple aerial refuelings, they could also undertake the long-distance missions formerly reserved for bombers. They had become viable if limited strategic platforms, and as such a threat to heavy bombers on the model of the B-17, B-29, and B-52.

As distinctions between strategic and tactical aircraft blurred, bomber advocates began to lose more budgetary battles than they won. The Air Force steered a growing share of its acquisition dollars to planes such as the F-15, which McDonnell Douglas began to design in 1966 as the limitations of the Century Series grew apparent, and in the mid-1990s came down in favor of the F-22 over building more B-2s.[37] Still, if the Air Force's preferred planes became smaller and lighter, the service's preferred mission remained the same: attacks on "high-value" targets in rear areas, rather than covering fire for troops on the ground. Hence the "oft-quoted slogan for the F-15, designed as an air superiority fighter, was: 'not one pound for air to ground.'"[38] Those capabilities were added later.

In Vietnam, the Air Force had to relearn an old lesson: air superiority required dogfighting capabilities and rapid-firing guns. It also required pilot skills that had atrophied since Korea. Even flying F-4s, Air Force pilots struggled in the Vietnam War. In the early years, the Air Force and Navy alike lost one plane on average for every two of the enemy they managed to shoot down. The Navy revamped its tactics and training—this was when the Navy created its "top gun" school—and managed to increase its kill ratio to 12:1, a higher figure than achieved in Korea. The Air Force did not seek a similar remedy until the war was nearly over, in 1973, and continued to lose planes at a higher rate.[39] By the time of the 1991 Gulf War, however, Navy pilots found their equipment and doctrine inferior to those of the Air Force. The Navy took advantage of the Reagan buildup to seek a 600-ship fleet. Rather than upgrading naval aviation, the admirals prepared to fight a (nonexistent) Soviet blue-water fleet.[40] The Air Force had failed to prepare for the sort of war that the nation asked it to fight in Vietnam. The Navy subsequently neglected to plan for the sort of war the nation asked it to fight in the Persian Gulf.

The Requirements Straightjacket:
The M16 and B-1B

DOD and the services sometimes ask system designers to bend if not break the laws of nature. That was the case when the Army set out to develop an assault rifle to replace the M1 of World War II. The Ordnance Department had spent nearly two decades developing the M1. The M16 took even longer. Why more than 20 years just for a rifle? Because of conflicts within the Army; between the Army and the Air Force and Marines, which rely on Army Ordnance for small arms; between the services and OSD; between the federal government and firms that manufactured arms and ammunition; between the United States and its allies within the North Atlantic Treaty Organization (NATO), which was seeking to standardize on ammunition.[41] Years were wasted on pointlessly meandering efforts including extensive tests based on near-irrelevant criteria such as accuracy at minus 65 degrees Fahrenheit and penetrating power at ranges far beyond those seen in combat. The conflicts drew in high-ranking generals, the White House, and both regular and specially constituted review boards and panels. They led to investigations by the Army's inspector general and by a congressional subcommittee. Incompatible technical requirements underlay what in hindsight and on the surface seems like bureaucratic farce. The main cause of the difficulties and delays? The Army in effect declined to accept the dictates of high school physics, imposing requirements that created artificial hurdles having little or no connection with known patterns of actual military usage.

In the beginning, the Army wanted a lightweight automatic weapon, one that would be simple, reliable, and not too costly. It was to fire the Army's then-standard .30 caliber ammunition and be built with existing production equipment. The .30 caliber round was heavy and powerful. Any rifle designed to fire it would have to be heavy too, like the M1, in order to absorb the recoil. Even the heaviest prototypes of the new assault rifle proved impossible to control in automatic firing. Already overburdened foot soldiers could not, in any case, have carried enough .30 caliber ammunition to do much automatic firing. Only when the Army grudgingly accepted lighter, less-powerful .22 caliber ammunition did it get, after many other difficulties, a more or less acceptable package of attributes in the M16.

Literally for years the Army was unable to acknowledge and confront the tradeoff between weight and recoil, a refusal that left the program virtually paralyzed. If the Army could not accept sensible tradeoffs in the requirements for something as uncomplicated as an assault rifle, it should be no surprise that the military sometimes insists on unrealistic requirements for complex systems incorporating exotic technologies that are harder for decision-makers to understand, the more so when advocates bill those technologies as avoiding the need for such tradeoffs.

The initial requirements for what became the B-1B bomber, set down in the 1960s, called for high-altitude flight at more than twice the speed of sound, Mach 2.2, coupled with the ability to fly a terrain-following course at treetop level beneath Soviet radar.[42] The genesis of the B-1 predates this set of requirements by some years, going back to the B-70, proposed in the 1950s as a replacement for the subsonic B-52. The B-70 was to reach Mach 3, fly above the reach of Soviet air defenses at 70,000 feet, and have a range of 7,000 nautical miles to enable round trips to targets deep inside Soviet Russia. President Eisenhower tried to kill the B-70 and President Kennedy succeeded, although two prototypes were built. The idea of a high-speed, high-altitude penetrating bomber resurfaced as the B-1 during the Nixon administration. President Jimmy Carter tried to kill the B-1. His success was temporary; the Air Force eventually purchased 100 B-1Bs, one of the fruits of the Reagan buildup.[43] The Reagan White House justified the B-1B on the basis that the planes could be placed in service earlier than stealthy B-2s, under development but not yet ready for production, supplementing elderly B-52s and heading off a "bomber gap." (All three planes had the same fundamental mission: delivery of nuclear warheads to targets in the USSR from bases in the United States.)

The critical requirements for the B-1, those that dictated its overall design, included a mission profile envisioned as high-altitude supersonic penetration followed by low-altitude bombing runs. The aborted B-70 had been designed to fly faster than Soviet fighters and above the presumed ceiling of Soviet antiaircraft missiles. Perhaps because in 1960 the Soviets had finally managed to bring down a high-flying U-2 spy plane (after many earlier failures), the B-1B was intended to instead hide from air defenses by approaching its target area at very low altitude, a few hundred feet or less. Airborne "look down, shoot down" radars were well in the future. Ground-based radars would be handicapped by a very short horizon and any returns they did pick up were likely to be masked by ground clutter. Surface-to-air missiles could not be brought to bear, again because of the B-1B's very low altitude. Nevertheless, the Air Force continued to insist on high-altitude supersonic flight, though never presenting a cogent rationale for the carry-over requirement, which is in fundamental conflict with the dynamic responsiveness needed for low-altitude terrain-following. To achieve high-altitude supersonic capability together with low-altitude maneuverability, the B-1B incorporates a moveable wing that sweeps back at high speeds to cut drag. The bulky, heavy (and costly) wing carry-through structure seriously compromised the B-1B's entire design, including the plane's low-altitude bombing mission: fully loaded B-1Bs proved unable to maneuver as intended during low-attitude penetration.

Along with questionable flight stability, the B-1B suffered technical troubles ranging from leaky fuel-carrying wings to faulty engine compressor blades and obsolete electronic countermeasures incapable of jamming Soviet radars.[44] Complex systems such as the B-1B always need engineering changes after entering service. Some postproduction modifications to the B-1B plainly fall in the category of normal fixes. That seems a reasonable characterization of the fuel leaks, a consequence of sealing the wings to avoid the added weight of internal bladders. But flight limitations resulting from the conflicting requirements of supersonic flight and low-altitude maneuverability stem directly from overambitious and inflexible requirements. These forced a cascading sequence of design decisions that ended up limiting the B-1B's mission capabilities. The requirements were unnecessary; the consequences included $30 billion in expenditures that might have gone elsewhere. No B-1Bs bombed Saddam Hussein's forces in 1991.

Examples of inappropriate requirements—arbitrary and unrealistic, unstable and constantly changing, and often both—can be cited almost endlessly. For the C-5A cargo plane,

> The original requirement . . . was "to be able to land on unprepared fields," and accompanying pictures showed tanks firing as they came out of the rear of the C-5A as though the enemy were only hundreds of yards away. No one would permit a C-5A to get anywhere near actual combat: its too costly, too easily damaged, and certainly too valuable. It belongs on the runways of major overseas airports . . . . But how much less costly would it have been if the unprepared field landing capability had been properly thought out and then thrown out early as it should have been?[45]

Like the B-1B, the C-5A was built and does fly. The extreme cases of unsuccessful programs are those such as the Navy's A-12, canceled long before flight tests could be contemplated. Although withheld from the 1991 Gulf War, B-1Bs participated in bombing campaigns in Serbia and Kosovo in 1999, in Afghanistan in 2001, and in Iraq in 2003. Even so, the Air Force has already retired nearly one-third of its B-1Bs, and would have retired all of them if Congress had not insisted that some remain in service.

As Polaris and aircraft programs such as the F-117 and F-16 show, the Pentagon has had its successes too. All three began with a reasonable set of requirements. For the F-16, these derived from the "lightweight fighter" program of the early 1970s, which itself had begun with an abbreviated request for proposals limited to 10 pages of technical discussion and limiting contractor responses in turn to 60 pages because of "horror stories involving truckloads of proposals being delivered for source-selection evaluation boards."[46] The lightweight fighter and F-16 programs figured prominently in the 1970s defense reform debate, have been much lauded since for bureaucratic streamlining, and continue to be put forward as models of enlightened Pentagon management.[47]

Measured against initial cost and schedule targets, the performance of DOD's major programs has improved since the 1960s. One reason is simply that the Pentagon has reined in some of the optimism that the services rely on to sell their more ambitious undertakings, recognizing that too many unmet promises create lasting political damage. This, of course, does nothing to control overall spending or shorten actual development times, which have continued to increase. And if schedule overruns have been reduced somewhat since the 1960s, they are still large, while cost increases over the original estimates average around 70 percent.[48] Thus the fundamental dilemma of acquisition remains: How can government best manage technically demanding projects that go on for years if not decades at enormous cost to the treasury, eventually resulting in systems with life-and-death implications for military personnel, perhaps for civilians, in the extreme for entire societies?

CHAPTER 7

RESEARCH AND WEAPONS DESIGN

A perfect ship of war is a desideratum which has never yet been attained . . . . Any near approach to perfection in one direction inevitably brings with it disadvantages in another.
—Committee on Designs for Ships of War,
UK Board of Admiralty, 1871[1]

A basic argument of this book is that technical requirements imposed by the armed forces constrain major design decisions excessively, that this is unnecessary, a prime cause of high costs, and that it sometimes undermines the very performance goals sought by the military. The essence of engineering design is the management of tradeoffs and compromise among competing objectives. Compromise does not imply providing the men and women of the American military with anything less than the best possible systems and equipment; effective management of tradeoffs is part of determining and achieving what is in fact "best."

The Army once asked for a "universal tractor" with a mean time between failures (MTBF) of 2,000 hours.[2] Such a requirement, the equivalent of a working year of eight-hour days without a mechanical failure, is absurd on its face. MTBFs for Caterpillar's D-7 bulldozer, a workhorse in use around the world, fall in the range of 60–80 hours. An Army requirement of even 200 hours would have been difficult to achieve and would in all likelihood have led to a "gold-plated" design of the sort the services have so often been criticized for seeking. In a final bit of foolishness, the Army wanted to be able to drop its new tractors by parachute.

As the tractor example illustrates, requirements too often represent wishful thinking. Too rarely do they reflect demonstrated military needs. As we saw for the M16 rifle and B-1 bomber, the services are often reluctant to compromise once requirements have been set. In other cases they keep changing the requirements. Either way, efforts to satisfy them drive up the first costs of acquisition and generally life-cycle costs as well, while the performance actually realized may also suffer. Yet the services have never trusted the judgment of either defense contractors or civil servants; in their view, only military professionals can understand what is needed for warfighting. Too often as a result, the services end up asking contractors for the impossible or the barely possible. Settling for the achievable is a choice which, if made at all, is typically a last resort and affects program outcomes only at the margins. Once overall system architecture has been locked in, which occurs early in design and development, further technical changes, those that can be accommodated within the existing architecture, rarely yield more than small capability improvements or cost reductions. If such changes are insufficient to solve technical problems encountered in downstream engineering development, the government, after many years and many billions of dollars, may have little choice but to accept a second-best system or else abandon the program, in which case the Department of Defense (DOD) may then be forced to start over, as with the Tri-Service Standoff Attack Missile mentioned in chapter 2.

The world of technology is a world of multiple solutions to poorly understood problems. Nowhere is that more true than in defense. This chapter explores the underlying reasons, beginning with a brief introduction to technological innovation itself. The next chapter explores the ramifications of technological and organizational complexity. Chapters 9 and 10 delve more deeply into the differences between military and commercial innovation.

Innovation and Diffusion

Technology evolves through processes resembling those that paleontologists call punctuated equilibrium. Radical innovations, relatively rare, interrupt and disrupt what would otherwise appear as a continuous flow of small, incremental changes: an increase by a percentage point or two in the efficiency of a jet engine's compressor stage; an improvement of a few percentage points in the production yield of a fabrication line for integrated circuits (ICs); a reduction from 13 to 9 parts in an automobile door assembly. Amidst this sea of incremental change, radical innovations—the jet (or gas turbine) engine itself, the microprocessor, "lean production" in the auto industry—stand out like rogue swells amidst an ocean of wavelets.[3]

Radical innovations sometimes begin with a discrete, identifiable invention or discovery—the synthesis of nylon at Du Pont in 1934, the demonstration of a working laser at Hughes Aircraft in 1960—but need not. They may also emerge more or less spontaneously through the accretion of incremental change, with no proximate act of invention. The Internet is the recent example par excellence. As a system, it combines many innovations, few of which had much visibility beyond the technical communities involved and the agencies that sponsored the work. Only after this long prehistory of incremental developments, which begins in the 1960s under DOD sponsorship, did the Internet burst into public consciousness. Similarly, a lengthy sequence of incremental improvements is the usual prelude to market introduction, as it was for nylon.[4] For commercial innovations, many years of slowly increasing sales may be followed, for the most successful innovations, by rapid "takeoff" (figure 6). During the period of slow diffusion and gathering momentum, innovating firms seek to raise performance and reduce costs.

An innovation may itself be counted as radical yet have its greatest impacts as part of some other, more complicated system. The semiconductor diode laser, a pathbreaking invention in its own right, when combined with low-loss optical fibers—an independent development—made possible fiber-optic communications links. Without these two separate streams of development, neither of them related in any direct way to computers, the Internet could not have emerged in anything like its present form.

Radical innovations, finally, even in this age of science, do not always or necessarily begin with research. While the initial synthesis of nylon and the demonstration of a working laser each marked the culmination of a

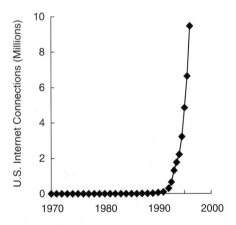

Figure 6   Early Growth of the Internet

Source: Based on data from <www.cyberatlas.com>.

lengthy process of scientific exploration, the microprocessor was the outcome of a pure exercise in engineering design. A Japanese firm, Busicom, had approached the semiconductor manufacturer Intel for a set of custom IC chips around which to build a simple, four-function calculator. Intel engineers proposed as an alternative a "universal" chip that could be programmed to implement the calculator functions (addition, subtraction, multiplication, division). The idea of a programmable chip was an old one, but had not previously been reduced to practice. It was the coming together of demand (by Busicom) and supply (a workable design concept for a computer-on-a-chip) that led to Intel's introduction, in 1971, of its 4004, the first commercially available microprocessor. Although Intel drew on a large existing knowledge base, no new research underlay the 4004.[5]

The explosive growth of the Internet in the 1990s so evident in figure 6 came as a surprise. Sometimes the surprise lies, not in how swiftly a new technology spreads, but in how slowly. For decades, numerically controlled (NC) machine tools—a form of automation in which digital processors guide milling cutters or other metal-working equipment—sold in numbers small enough to puzzle analysts and disappoint the machine tool manufacturers that brought NC equipment to market.

Numerical control originated in Air Force-sponsored R&D conducted at MIT during the late 1940s. The Air Force wanted better methods for machining large parts such as helicopter blades and wing spars, which were costly and time consuming to make and might be ruined when nearly complete by machining errors. Early NC tools were difficult to use and expensive to maintain, so much so that in 1958 the Air Force purchased more than 100 five-axis milling machines for installation in the plants of contractors that declined to invest their own funds. Simpler and less expensive NC equipment followed, but in 1968 fewer than 14,000 NC machines were at work in the United States. The numbers mounted, reaching about 100,000 in the early 1980s, by which time the technology was some 25 years old. Yet even this total represented only 5–6 percent of installed machine tools. Some observers found explanations in the defense origins of NC, blamed for needless complexity, others in a backward U.S. machine tool industry, said to have done a poor job of incorporating NC into its product offerings, and still others among prospective customers, claimed by those who believed NC self-evidently cost-effective to be oblivious to their own self-interest. Better explanations center on the practical difficulties of learning by users, with customers slow and cautious in exploring what NC, a considerable investment for a small shop, could do. As the lessons of experience spread, diffusion gradually picked up speed.[6]

A similar dynamic, though with a different set of causes, underlay the slow rate at which "organizational technologies" such as statistical quality control, total quality management, and continuous improvement spread within U.S. industry during the 1980s and 1990s. It took American firms, large and small, including those under severe competitive pressure from rivals in Japan and elsewhere that had made quality a centerpiece of business strategy, some 15 years to understand and adopt comparable methods. The reasons included difficulty in untangling myth and misinformation from methods and practices of demonstrated effectiveness.[7]

For the Internet, as for the microprocessor, many of the end-applications that eventually took on great importance were not at first apparent. In the case of jet propulsion, the uncertainties lay more in the pathway than the end-point. Although gas turbines operate on a thermodynamic cycle patented in 1872, and industrial turbines (of very low efficiency) were built early in the twentieth century, engines suitable for aircraft propulsion could not be built until the 1930s because component technologies were inadequate. Reasonable power output from a gas turbine engine of reasonable size and weight requires reasonably efficient compressor and turbine stages; until this could be achieved, nearly all of the turbine's output would be absorbed in driving the compressor, leaving nothing for propulsion. Other obstacles included the design of burners that would maintain stable combustion over a range of pressures and flow velocities and materials, for the turbine especially, with useful strength at high temperatures. The endpoint was plain enough: high power at high altitude, where piston engines gasped for air. To reach it required successively better solutions to a series of large independent technical problems, related mostly in that failure to solve any one of them threatened to bar attainment of a satisfactory overall system.

Jet-propelled planes flew on an experimental basis in Germany in 1939, two years later in England, and in the United States, following transfer of technology by America's ally, Britain, in 1942. Early jets had limitations including staggering rates of fuel consumption. In 1948, Secretary of Defense James V. Forrestal explained to a congressional committee "that the fuel consumption of a jet Air Force was approximately forty-five times that of a traditional air force . . . ."[8] Lockheed's P-80 (later F-80) Shooting Star, which went into production in 1944 (none saw combat until the Korean War), burned so much fuel that its operating radius was little more than 100 miles. This could be extended to about 250 miles with external drop tanks, but the pilot had to jettison them before an engagement. Judged on combat radius, the F-80 was far inferior to the piston-engined

Table 6   Jet Engine Performance Improvements, World War II–Present

	World War II (ca. 1945)	*Present (ca. 2005)*
Maximum thrust (pounds)	1500–2000	> 100,000
Fuel consumption (pounds of fuel per pound of thrust per hour)	1–1.5	< 0.4
Thrust-to-weight ratio	~ 1.5	> 8
Time between overhauls (hours)	25–100	> 30,000
Maximum speed, military fighter (miles per hour)	500 +	~ 1500

P-51 Mustang of World War II, which had a range approaching 2000 miles when fitted with the largest available drop tanks.

Over the years, jet-propelled planes gained in range and grew bigger, heavier, faster, and more reliable as the result of a long series of incremental improvements to jet engines themselves (table 6), coupled with aerodynamic developments such as area–rule fuselage designs (it is the "area rule" that prescribes a "Coke-bottle" shape for supersonic aircraft). Jet and gas turbine powerplants moved from fighters to bombers and then to commercial airliners, helicopters, cruise missiles, warships, and eventually to the Army's M1 Abrams tank. Along the way, commercial applications benefited from DOD spending on R&D and procurement.

Gains in functional performance, usually accompanied by reductions in cost, pace diffusion of technological innovations. As will be discussed in chapter 9, many of the improvements stem from learning associated with operating experience and fed back into design and manufacturing. Meanwhile, potential users, those who have not yet adopted the innovation, must learn enough to decide whether to invest and, having done so, how to make effective use of what they have purchased (or, for organizational innovations, adopted). In the case of the Internet, learning proceeded rapidly once the World Wide Web became operational and the browser Mosaic had been introduced. For NC machine tools, learning was slow, despite technical improvements. Surveys revealed that many potential purchasers of NC equipment, small firms especially, did not understand how NC might be applied to their business and, when quick returns on investment might be possible, did not realize it.[9]

As both the Internet and NC examples suggest, diffusion of innovations requires that large numbers of people and organizations, variously welcoming or resistant to change, learn how a new technology works and what it might do for them. This sort of learning is necessarily experiential, often a matter of trial-and-error and trial-and-success. When adoption calls for major changes in organizational structure and work routines—or, in the military, doctrine—diffusion and deployment are likely to be drawn-out and disruptive. That was a contributing factor in the NC case, not least in firms where skilled machinists feared for their jobs. Other firms purchased NC equipment but did not train their employees to use it. Innovations that spur major shifts in business strategy and structure may drive some firms into bankruptcy, in the extreme obliterate entire industries. While prospective winners may welcome such innovations, those threatened quite naturally resist.

In the civilian economy, incremental technical change is everywhere. Diesel engines for pickup trucks get more power and better fuel economy thanks to higher-pressure fuel injection and stiffer chassis members thanks to hydro-formed frame rails. Hardly anyone notices except employees of the innovating firms and the most technically attuned of their customers. The number of circuit elements on IC chips, now many millions, doubles every 18 months or so, as depicted by Moore's Law; what most people notice is that the prices of PCs and other consumer electronics products fall even as their capabilities increase. In the world of defense, by contrast, lengthy acquisition cycles lead to discontinuous innovation: big jumps in performance spaced many years apart.

The Navy's long-running efforts to develop rapid-fire guns and surface-to-air missiles to protect its ships from enemy planes illustrate how system-level innovations commonly evolve. This example also shows the continuing prominence of trial-and-error despite rapid advances in science and engineering science. In the late stages of World War II, Japanese kamikaze attacks threatened to overwhelm the U.S. Navy's defenses. Obsolete aircraft otherwise useless in combat, even trainers, could be stuffed with explosives and crashed into a ship by unskilled pilots. Though sluggish and slow, these human-guided cruise missiles managed to hit and damage the Navy's ships with much greater frequency than dive bombers and torpedo planes making conventional attacks.[10]

After the war, it was evident that future enemies might at some point acquire nuclear warheads with the power to obliterate a warship able to withstand multiple high-explosive bombs or torpedoes. Defensive systems would have little time to work and a near miss would probably be as devastating as a hit. Incoming planes or missiles would have to be engaged at far greater ranges than those of World War II and shot down with far higher success rates than the Navy had achieved against kamikaze attacks. Even with the best radars of the time, high-altitude bombers were difficult to spot. Missiles, smaller and potentially swifter, promised to be still harder to track. No matter how good the air cover provided by the Navy's own fighters, some attackers would get through. Shipboard defenses would have to down them at ranges of 5 or 10 miles or else divert them using counter-measures.

The Navy and its contractors began by attempting to extend the range of antiaircraft fire, designing and testing many different types of rapid-fire guns.[11] Some were extensions of conventional technologies. Others used liquid propellants or rocket assistance to increase projectile velocity and range. These approaches came to nothing. The Navy then declared the antiaircraft gun obsolete, accelerated R&D on radar-guided surface-to-air missiles, and began to design ships that bristled with missiles while forego-ing guns even for close-in protection. The new surface-to-air missiles did not work well either. In one sea trial, "Of the 178 raiders that evaded carrier fighters, the ships' missiles downed, aside from 'numerous friendly and non-exercise aircraft,' only twelve of the enemy. Antiaircraft gunfire accounted for three more."[12] After witnessing a 1962 trial in which three successive missiles failed to hit a slow-moving propeller-driven drone flying a predictable course, President John F. Kennedy ordered the Navy to add guns to the missile cruisers it had begun to build. Shortly afterward, the Secretary of the Navy initiated a "get well" program for surface-to-air missiles. It would run for a dozen years. In the middle 1960s, the Navy began work on the computer-controlled Aegis system, able to "perform search, track and missile guidance functions simultaneously with a track capacity of over 100 targets."[13] Nearly two decades would pass before Aegis became operational.[14]

This story is not atypical. Much military technological development culminates in neither quick and notable success, such as the Navy achieved with nuclear submarines, nor dismal failure. Instead, development proceeds through a lengthy sequence of partial successes and partial failures. Some of the lessons of these successes and failures are absorbed; others may be over-looked or ignored.

Military leaders have frequently been portrayed as conservative if not obtuse, stubbornly unwilling to adopt new technologies and adapt doctrine to them. That is not entirely fair. Armies did cling to horse cavalry and sabers well into the age of the machine gun; even as World War II approached, the U.S. Army required Air Corps pilots to supplement their advanced flight training courses with practice in horsemanship. But these are stock examples. Military men in the nineteenth century had been alert to the potential of mass-produced small arms for equipping mass armies and railroads for moving them for attack or defense. Navies quite literally raced to embrace steam and steel and later explored and debated the strengths and vulnerabilities of aircraft carriers and battleships. By the beginning of

World War II, it would be wrong, as a generalization, to accuse militaries of being reluctant innovators. Sometimes they were; sometimes they were not. As we will see in chapter 9, the U.S. Navy, thanks to well-considered trials and experimentation during the 1920s and 1930s, came to a reasonably accurate understanding of what air power at sea could accomplish. Great Britain, albeit under a different set of political and budgetary constraints, misread much the same evidence, entering World War II handicapped by inferior aircraft carriers and carrier-borne planes. Technical enthusiasts in prewar militaries pressed optimistic proposals and made unrealistic demands on contractors, just as they would do in later decades.

The overall picture, then, was mixed if not muddled, at least through World War II. The U.S. Navy made generally good choices in aviation but found itself unprepared for night battles in the Pacific, engagements for which Japan's navy had trained extensively. The U.S. Army deceived itself badly concerning tanks, as we saw in chapter 4, but chose correctly in standardizing on a relatively small number of aircraft designs. Those were reasonably well suited to their assigned tasks. It was just that the tasks had not been appropriately selected and defined.

Now as then, the uncertainties, costs, and risks of innovation push militaries into postures of at least moderate conservatism. No one can know what war might bring: the opponent; the location of critical engagements, much less conditions of weather and terrain. Foresight will always be flawed and mistakes can be terribly costly. Since all military professionals know this, they do not lightly abandon demonstrated capabilities in favor of the new and unproven. At the same time, they understand the dangers of excessive caution: if the enemy has a superior weapon, you had better be prepared to counter it, one way or another.

While uncertainty is pervasive in innovation, military and commercial alike, it takes different forms and the uncertainties are more difficult to resolve in military affairs. The microprocessor and PC illustrate common patterns in the civilian economy: high levels of initial uncertainty that soon decline as the lessons of experience become known and circulate through discussion and debate in forums including the business and trade press and technical meetings, and even through advertising.

Like entrepreneurs in the civilian economy, military professionals need few reminders of uncertainty. Writings on strategy are full of admonitions on the subject. The fundamental theme: commanding officers and their staffs must prepare for battle with painstaking attention to detail while recognizing that matters never go as expected. Any plan will have to be altered, perhaps drastically, as events unfold. To this age-old understanding of contingency, new technologies add new unknowns. Will infantrymen expose themselves to enemy fire long enough to steer wire- or laser-guided antitank missiles to their targets? There may be no way to answer such questions in advance. After all, military analysts argued for years over the fraction of riflemen who actually fired their weapons during World War II, and of those who did the fraction who aimed at anything.

Compared with commercial innovation, experience accumulates slowly in military affairs. Information flows are intermittent rather than continuous, lessons held closely, perhaps kept secret, for reasons of national security or personal and organizational self-interest. Uncertainty contributes to conservatism, increasing the likelihood that existing weapons, such as the battleship, will be retained well after they have been superseded, if for no other reason than insurance, and that innovations such as strategic bombing, once adopted, will be pursued until overwhelming evidence of shortcomings dictates alterations to technology or doctrine or both.

Research, Development, and Design

Technological systems, both military and civilian, embody multiple design tradeoffs or compromises. For an airplane, high speed, high rate of climb, and short take-off distance require high power or, for a jet engine, thrust. More thrust means more fuel consumed, especially if purchased with afterburning jet engines, which are notoriously inefficient. Carrying more fuel to enlarge the combat radius adds weight, requiring more power to achieve the desired rate-of-climb and top speed. The laws of fluid mechanics dictate that higher lift means higher drag, meaning that aerodynamic characteristics suited to high-speed flight (i.e., low drag) differ substantially from those needed for maneuverability and also for short take-off and landing (high lift). Alternatives such as the B-1B's pivoting wings bring their own penalties in terms of cost, weight, bulk, and maintenance. Many of the relationships that govern these tradeoffs are nonlinear. Stealth comes at the expense of flight stability and designers have little latitude for compromise because a small increase in radar cross section (RCS) makes the plane visible to radar at much greater distances: detection range increases as the fourth power of RCS. This means that, in theory, an F-117 should be able to fly undetected a tenth of a mile from an air defense radar that can spot a semistealthy F/A-18E/F at 100 miles. The difference is that the F-117 has the RCS of a golf ball (if it were made of aluminum), the F/A-18E/F that of a 3-foot sphere.

Submarines illustrate a more complicated set of technically imposed tradeoffs. Diesel-electric submarines are quieter than nuclear-powered submarines, which gives them underwater stealth at the price of two separate propulsion systems: an air-breathing engine for surface cruising and charging the massive banks of storage batteries that drive the electric motors when submerged. Because the physical laws governing drag differ between surface and underwater travel (there is no underwater counterpart to surface wake), hull shape can be optimized for speed on the surface or underwater but not both. Yet both surface and underwater speed are tactically critical. Given typical design choices, a diesel-electric submarine cannot travel fast enough on the surface to catch any but sluggish merchant vessels. It cannot escape from enemy warships or keep up with a friendly fleet. A diesel-electric submarine has no alternative for attack but to lie in wait or maneuver to intercept on a closing course. Slow underwater speed, limited by electrical power, means the submarine can only hide, not run from a surface ship that detects it via sonar. In essence, nuclear power enabled designers to overcome underwater drag through brute force (early nuclear submarines could outrun then-standard torpedoes) and extend underwater range almost indefinitely, erasing at a stroke the major shortcomings of submarines as war machines. Easier detection by enemy sonar, a consequence of greater powerplant and hydrodynamic noise, was an inconsequential penalty.

Design decisions occur as nested sets. Each has consequences, feeding forward to shape and constrain future choices and in some cases also feeding back to force reconsideration of earlier decisions. The final configuration of a jet engine reflects tradeoffs at both the system level (e.g., pressure ratio across the compressor stage) and in the details of subsystems and components (tip seals to reduce leakage past the compressor blades). Integrating engine and airframe means further tradeoffs, such as variable-geometry air inlets. Today, designers seeking a low RCS must also take steps to minimize reflections from inlet ducts and the spinning blades on the compressor within, steps that may lead to substantial penalties in available thrust.

The Two Meanings of Design and the Two Meanings of Development

Research, development, and design are easily confused. Research refers to the search for new knowledge, often guided by scientific theory. The goal is discovery. Development and design are broader terms. Their goal is a working system. Both design and development, furthermore, have two meanings, related but distinct.

The first meaning of design is associated with conceptualization and preliminary design, the exploration of alternative architectures and the selection of one of these alternatives for the final system. The second meaning applies to the definition of details, all of which must be fully specified so that production processes create identical systems. The first meaning of development refers to extensions of research as part of the exploration of new domains of knowledge—hence "R&D." Sometimes but not always, a particular end product may be in view. Alternatively, development can mean engineering development, associated with design rather than research and referring to the iterative process of analysis, testing, redesign, and refinement through which an end product or system takes on its final shape. Thus "design and development" or "D&D" can be usefully distinguished from R&D.[15] D&D begins with assessment of alternative concepts, based heavily on intuition and experience, and culminates in full definition of details, traditionally specified in the form of blueprints and process sheets and now often as digital data.

Practitioners understand these distinctions. Policymakers may not. Conceptual design and detail design are two ends of a continuum. Decisions at the conceptual end determine the overall configuration; decisions at the detail end determine discrete features necessary to fully realize the overall design. The two meanings of development, by contrast, are only loosely related. In the first sense, as an extension of research, development tasks include verifying and validating experimental results and theoretical predictions, exploring particular cases, and determining the accuracy and limits of mathematical models. This sort of work adds to the knowledge base of technology and science but has no necessary connection to particular end products or systems. When linked with design rather than research, development refers to the refinement of a concretely conceived product or system, work that often continues even after systems have reached the field (as for the B-1B). The process as a whole may be called engineering development, product (or system) development, or product engineering (or process development, software development, and so on). As a technical activity, it has little in common with research even though uncertainty is omnipresent in both D&D and R&D and surprise common, while evaluative criteria may be elusive and subject to debate.

Sometimes research becomes the starting point for military innovation. That was the case with stealth, which began in the early 1950s with experiments on scale models to explore how radar returns varied with the geometry of reflecting surfaces.[16] Over the next quarter-century, as stealth advanced from those first laboratory studies to the Have Blue demonstrators, research fed a generic (though secret) technology base. The Have Blue program confirmed and extended earlier work based on mathematical analysis and small-scale tests and provided a basis for designing the F-117 and B-2. In other cases, including the M16 rifle and Intel's first microprocessor, the essential task from the beginning was one of engineering design—D&D

without R&D. The difference is that the microprocessor qualifies as an invention—Intel's 4004 was the first of a fundamentally new class of electronic devices—while the M16 was just another assault weapon.

Product Cycles

Design and development cycles have shortened over the past several decades in much of the civilian economy. Firms in highly competitive industries have telescoped the work preceding new product introductions. In the auto industry, for example, tasks that once took five or six years (i.e., from concept selection to production startup) are completed in half that time or less. Firms use both managerial tools, such as conducting tasks in parallel rather than sequentially, and technical tools, such as computer-assisted engineering (CAE), which save time by automating routine tasks ranging from drafting to complicated calculations based on mathematical models.

While product cycles have become shorter in the civilian economy, they have grown longer in defense. According to the Defense Science Board (DSB), Air Force programs doubled in average duration between the early 1970s and the late 1990s. Army programs did not change much and Navy programs grew longer, but by less than those of the Air Force. On the measure chosen by DSB—time from formal program start-up to initial operational capability—programs run by all three services appeared to be converging to about 10 years on average.[17] Because this measure excludes preprocurement steps in the usual research, development, test, and evaluation (RDT&E) sequence such as exploratory research and proof-of-principle demonstration, more inclusive measures would extend acquisition cycles by several to many years.

In the civilian economy, most new products and systems represent modest conceptual and technological advances over their predecessors. Moore's Law, for example, has held for more than four decades precisely because the primary dynamic has been steady incremental improvement. (That the relationship exists demonstrates the point: on a Moore's Law–like plot, a radical innovation would appear as a discontinuity, a jump from one trend line to another.) Military systems and equipment are far more likely to be all-new designs that incorporate major breaks with past technologies, as illustrated by the enormously complex Aegis air-defense system. They then remain in service much longer than typical commercial products. The last of the 744 B-52s built by Boeing left the assembly line in 1962 and may still be flying in 2050. Quite understandably, whenever money becomes available for a new program, the armed forces seek the most capable system possible to fight the most powerful opponents imaginable decades into the future. The general approach has often been justified by arguing that these very sophisticated weapons will more-or-less automatically be able to handle whatever lesser threats may appear. The Air Force experience with its F-105s in Vietnam points to the fallacy, one to which the concluding chapter will return.

Preliminary Design Decisions

Design is a fluid, open-ended activity, synthesis not analysis, with little basis in science and theory. Put differently, it is an art rather than a science (or an art based on science). This is a familiar point, as expressed by Elting Morison:

> If I understand the matter correctly, engineers must begin with the urge . . . to take separate, diverse elements and put them together into an organized situation or construct that will really work. This is the urge, if I understand it, of the artist, to give some concrete form to an organizing perception.[18]

For military and commercial systems alike, preliminary or conceptual design largely determines both functional performance and life-cycle costs.[19] This simple truth, illustrated for quite prosaic pieces of equipment by the M16 and for complex systems by the B-1B, is too often obscured by fascination with the bits and pieces of technology such as IC chips. When acquisition programs run into trouble, the reasons can usually be traced to preliminary design and through this to the requirements laid down by the military.

Examples of preliminary design decisions: settling on a shape that will minimize a plane's RCS; choosing between solar cells for spacecraft power and nuclear isotopes (or both, for redundancy); selecting the number of cylinders, valves, and camshafts in an auto engine. Some decisions may be more or less givens, imposed by convention or product differentiation (air-superiority fighters like the F-22 get two engines, luxury cars get engines with more cylinders). Some reflect national or corporate style (Soviet fighters traditionally had especially powerful engines, Volvos prominent safety features). "Design" also brings to mind rows of drafters and detailers sitting at desks and drawing boards—today, computer terminals—pouring over catalogues and handbooks, picking fasteners to hold the cowlings on the airplane's engines, determining the wire gauge for connecting the spacecraft power supply to its electrical bus, specifying the surface finish for the auto engine's cylinder bores. Although well removed from the worlds of

R&D and science, detail design is far from inconsequential. The cowling could fly off, the wires overheat and burn out. The location, surface finish, and inspection methods for each bolt and rivet hole in an airplane, many tens of thousands, must be specified because the great majority of aircraft structural failures start at fastener holes, generally at flaws left during fabrication, and every precaution must be taken to minimize flaws and detect those that do occur. A low RCS depends not only on overall shape; close attention must also be paid to panel joints, blend radii (e.g., between wings and fuselage), and sensor windows. Design details will be checked and reviewed many times over. Still, they are of a different nature than preliminary design decisions, because mistakes at the detail level, provided they are caught soon enough, can be rectified with only minor consequences for project budget and schedule. There is often no way to fix a poor conceptual design except to start over.

Within constraints imposed by the technological "state-of-the-art," conceptual design begins with a nearly blank slate. Existing knowledge from many sources comes together, perhaps supplemented by new knowledge from research, with the aid of rules-of-thumb, heuristics, and experience-based judgment. Rarely are there direct pathways to anything approaching an optimal design. The quantitative methods of CAE and the tools of engineering science more broadly can lead designers to the "best" configuration only for the simplest of artifacts: a spar, not a wing; a turbine blade, not a jet engine.[20] Almost all tools and methods are "one way," in the sense that predictive calculations begin with a design complete in many details and the mathematics cannot be run "backward" to yield prescriptions for improvement. As a result, designers and analysts must begin by establishing a baseline configuration, exercise their models to predict its behavior, then alter one or another parameter and run the model again to see what changes. Such an approach can be used, for example, to "tune" an airplane wing to reduce the possibility of flutter—potentially destructive vibrations excited by aerodynamic forces that can shake a plane to pieces—but even the most advanced methods cannot turn a poor wing design into a good one. For preliminary design, or big changes in design configuration, engineers have little recourse except to rely on judgment and experience.

One reason for these limitations is that engineering science, like most other sciences, follows a reductionist strategy—breaking things down to understand them—while design is a matter of putting things together, of synthesis rather than analysis. The two activities are fundamentally different. That is why the mathematical models of engineering science, no matter how accurate their predictions, provide little help in arriving at a satisfactory initial concept. Military and commercial design groups alike spend much time and effort on "trade studies"—e.g., CAE calculations in which parameters are systematically varied—but such work cannot even begin without a reasonably complete system configuration. Once the basic design has been laid down, architecture quickly becomes locked in because anything beyond relatively small changes is likely to require starting over, which is costly, time consuming, and potentially embarrassing. For military programs, system concepts and architecture are often fixed before the first formal program reviews, if only because there may be little to review until a preliminary design has been completed.

CAE-based predictions, finally, sometimes go badly wrong. Despite extensive mathematical analysis, a full-scale prototype wing for the Air Force's C-17 transport buckled during its "proof test" at a load well short of that it should have sustained (according to calculations), necessitating an expensive redesign. Why? Wing structures combine large numbers of ribs, spars, and skin panels. These share the applied loads. In this case, the CAE model was inadequate. It did not accurately represent load-sharing among the many structural elements. (If such models were fully reliable, proof tests, which are quite costly, would no longer be needed.) In another example, test pilots flying prototypes of the Navy's F/A-18E/F in 1996–1997 unexpectedly encountered "wing drop": in some kinds of turns, one wing stalled before the other, causing a sharp roll of 30 degrees or more. Computer modeling and simulation failed to reveal the causes. Nothing like this had occurred with earlier versions of the F/A-18 (which differed substantially). The eventual solution came, not from analysis or theory, but from experimental studies suggested by past experience with other aircraft.

Because engineers get little help from formal models and methods in making preliminary design decisions, they commonly rely (often unknowingly) on heuristics that have worked for them in the past. Over more than 40 years, planes from Lockheed's celebrated Skunk Works bore the stamp of Clarence "Kelly" Johnson. At the same time, Lockheed's planes were identifiably American: U.S. and Soviet aircraft differed systematically. During the 1950s, Russian design bureaus created jet fighters with unusually high thrust-to-weight ratios, in part to compensate for the aerodynamic drag caused by crude fabrication practices. Those in turn reflected lessons from World War II, when Soviet planes might not survive a week in combat and came to be viewed as expendable.[21] Because most military conscripts were poorly educated and unskilled, the design bureaus also followed practices intended to yield high reliability between scheduled overhauls, which were conducted at depots staffed by specially selected and trained workers. The U.S. military, which drew its recruits from a higher quality pool, carries out much more maintenance and repair in the field. These are examples of contextual factors that affect both preliminary and detail design decisions.

Testing

Until a design has been validated in the hands of users, development cannot be considered complete and performance remains uncertain. This simple proposition holds for chemical plants and computer software, C^3I (command, control, communications, and intelligence) and telephone networks, the F/A-18E/F and the space shuttle. In the civilian economy, validation begins with testing programs viewed by business firms as an integral part of development. Any firm competing in a market economy has strong incentives to subject new products to realistic but demanding tests that will identify shortcomings early and address them promptly. In defense, the incentives are quite different: to pass tests in order to safeguard the program from attack.

> Problems revealed in flight tests caused two programs . . . reviewed—the Theater High Altitude Area Defense system and the DarkStar unmanned aerial vehicle—to take twice as long to develop as planned.
>
> [S]everal failures in flight tests of the Theater High Altitude Area Defense system were traced to problems that could have been discovered in ground testing.
>
> [C]ommercial firms avoid skipping key events and holding hollow tests . . . .
>
> Leading commercial firms have learned [that profits are] threatened . . . if unknowns about a product are not resolved early, when costs are low and more options are available. The role of testing under these circumstances is constructive . . . . Success for a weapon system . . . . is influenced by the competition for funding and the quest for top performance . . . . Testing plays a less constructive role in DOD because a failure in a key test can jeopardize program support.[22]

Unlike firms developing products for commercial markets, defense contractors may be unable to treat developmental testing as a vehicle for technological learning. Not only must they meet the sometimes-arbitrary hurdles imposed by the acquisition bureaucracy, but because the appearance of shortcomings, real or apparent, can endanger the entire program, the first goal must always be to pass the tests, whatever they are, rather than use tests to improve the product.

Sometimes DOD's test and acceptance procedures do not even attempt realism. The Army's Divisional Air Defense antiaircraft gun (DIVAD, also known as Sergeant York) was originally intended to shoot down high-speed jet fighters. Over time, the requirements were relaxed to the point of downing hovering helicopters. When, after 10 years, DIVAD could not accomplish even this task with any consistency (to be fair, not an easy one if the helicopter pops up to launch a missile from several miles away), the Office of the Secretary of Defense canceled the program and the Army scrapped the 65 systems already delivered.[23]

Why did the Army act so differently in the case of the M16 rifle, giving it tests that could not be passed? Because the M16 was essential to replace the old M1, long obsolescent. Since foot soldiers had to have a new weapon, there was little likelihood that the M16 program would be canceled and bureaucratic infighting could flourish unchecked. The need for the Army's Bradley fighting vehicle, conversely, was constantly in question:

> The [Bradley] was modified many times during its 17-year development history. The key issue is whether this vehicle bears any resemblance to its original requirement and whether its high cost can be justified in terms of what it can do on the battlefield.[24]

Because the program was vulnerable, the Army gave the Bradley tests it could not fail: literally for years, the Army shunned life-fire trials, even though the Bradley's ineffective armor posed obvious dangers to soldiers riding inside.[25]

Whether for DIVAD, the Bradley, or missile defense, peacetime tests, trials, and exercises cannot adequately simulate warfighting, not "normal" combat and certainly not the "unknown unknowns" that a war of any duration and intensity invariably brings. Life-and-death implications, meanwhile, make stringent testing of military systems doubly difficult and doubly critical. If military systems fail to work as expected when used in anger, the repercussions can be far more serious than in the civilian economy, in the extreme nearly unimaginable (e.g., accidental nuclear war). Human ingenuity adds robustness and resilience to technical systems. Stop-gap fixes may be possible if a wide-area computer network crashes or the electric power grid goes down, buying time to diagnose the problem and find remedies. That will often be impossible in war. A system that performs poorly may have to be shelved or else a heavy price paid in casualties. One of the reasons the services insist on multiple systems with redundant or overlapping capabilities is to provide insurance against the prospect that peacetime testing will fail to predict wartime performance.

Army Tanks: Choice and Compromise

Armored vehicles illustrate the ways in which choices among conflicting objectives cascade. Since the first tanks were built during World War I, technological competition has revolved around protective armor, firepower, and mobility (speed, acceleration, range). Thicker armor adds weight, not only the weight of the armor itself but the stronger supporting structure, suspension, and tracks needed to support it. A heavier tank will be less agile, less able to dodge the enemy's fire, unless power is raised commensurately. Adding horsepower to maintain speed and acceleration means a heavier powerplant that will consume more fuel, which the tank must carry and the army's logistics organization must transport to the front (at a cost the U.S. Army estimated to be $600 per gallon for the 2003 invasion of Iraq). A bigger engine will also generate more heat, which, if it cannot be masked, means a stronger infrared signature for the enemy's antitank weapons to home in on.

When the U.S. Army chose a gas turbine powerplant for its 70-ton M1 Abrams main battle tank, it sacrificed fuel economy compared to the diesel engines installed in earlier tanks. Whether intended for tanks or aircraft, gas turbines are smaller and lighter but less efficient than piston engines. In effect, the Army opted for speed and acceleration at the expense of range. During the 1991 Gulf War, its tank columns sometimes had to slow or stop to wait for fuel convoys to catch up. In 2000, the Army began work on a more efficient gas turbine for the Abrams, the first step in a replacement program expected to cost around $3 billion.

Armor itself illustrates a superficially straightforward technology that remains well beyond the reach of engineering science. During high-velocity impacts, both projectiles and armor behave in quite bizarre fashion, exhibiting phenomena unlike those observed under ordinary conditions. As a result, standard laboratory methods provide little useful information. While live fire tests can determine whether a given thickness of a given armor will stop a given projectile at a given impact velocity and angle, even a large body of test results provides a poor basis for generalization in the absence of theory and mathematics. Such "first principles" understanding, on a microscale in the microseconds following impact, promises improvements in both armor and in antitank munitions and remains a research target. But proposals to supplement and possibly replace testing and experiment with computer modeling have been opposed by those who insist that mathematical models cannot as yet capture the complexities of actual phenomena and might prove dangerously misleading.

Armies transport their armored vehicles by truck (or rail or water) when they can. This means highway clearances limit overall width, a major concern during the Cold War because expected routes of Soviet advance into Western Europe included mountainous regions with narrow bridges and tunnels. With overall width constrained, choosing thicker armor will leave less space inside the tank's hull for crew, electronics, the powerplant, and its fuel. Ammunition takes space too; everything else the same, the bigger the main gun the fewer reloads a tank can take into battle. Early versions of the Abrams fitted with a 105 mm gun carried 55 rounds (normally divided among armor-piercing projectiles for fighting other tanks, high-explosive shells for attacking bunkers and troop concentrations, and several other specialized rounds); the later M1-A1, with a 120 mm cannon, must get by with 40.

Over the years, the arms race in tanks proceeded much like that in battleships. When one country adopted a more powerful gun, others had to strengthen their armor. Considered in the abstract, this sort of military competition might seem to resemble market-based commercial competition. But in commercial races, such as those involving IC chips, disk drives, or mobile telephone handsets, technical requirements are more malleable, subject to change if costs escalate or intractable engineering problems arise. Military system designers have less flexibility. During the Cold War, the Red Army fielded four or five times as many tanks as the U.S. Army.[26] Those of the United States had to be better.

The Limits of Technical Rationality

When the United States set out to develop nuclear-tipped intercontinental ballistic missiles (ICBMs) in the 1950s, the stakes were almost unimaginably high. Technical failures might cause or contribute to global cataclysm. Did this mean, once the decision to proceed with ICBM development had been taken, that technical logic guided major design decisions? Not at all. As we now know from painstaking retrospective examination, not only was technical rationality not determining, it could not have been determining. There were too many uncertainties, particularly in the early years, no clearly superior approach, a lack of formal methods for comparing alternatives.[27]

A long-running debate ensued among proponents of radio control from ground stations, star-sighting systems carried on board the missile, and inertial guidance. The last of these, in essence "dead reckoning," relies on continuous measurement of acceleration during the powered or boost stages of flight, when the missile can be steered. Based on measured acceleration, a small special-purpose computer on board the missile would calculate the instantaneous path and compare it to the desired trajectory, predetermined and loaded earlier into memory. The guidance system would then steer the missile so as to minimize the instantaneous difference between the inertially determined path and the desired trajectory. At the end of the boost phase, the missile would, if everything worked correctly, have the proper speed and direction to coast "ballistically" to its intended target.

Inertial guidance at first seemed more problematic than the other alternatives, yet eventually won out. MIT's Charles Stark Draper strongly advocated the method and the Air Force trusted him. Draper was an acknowledged expert, but the Air Force based its choice on more than his reputation. Draper and his colleagues at the MIT Instrumentation Laboratory had created graduate engineering courses and indeed entire programs of study for officers sent by the Air Force to MIT for advanced training. In essence, Draper taught them to conceptualize the problem of ICBM guidance as he himself did. Some of these officers went on to become key Pentagon decision-makers. Draper's deep faith in and strong advocacy of inertial guidance could not and cannot be rationalized. Even today, it is not clear which of the competing systems was or is "best."

As this and other examples, such as the F/A-18E/F's wing drop, show, many technology-related matters continue to depend on expert judgments. Such judgments are frequently in error, always to be questioned. When available, formal methods such as those of engineering science must be presumed superior. But almost by definition formal decision methods will not be available for critical issues (else those issues would not be identified as "critical") and the advice available to policymakers will rest on informal or tacit knowledge and expertise. More frequently than lay observers may wish to accept, there are no reliable technical methods to illuminate the issues of greatest import for national security.

Among the implications: funneling large sums of money into RDT&E is no guarantee of effective weapons systems. The requirements set by DOD and the services must also be appropriate, in two senses. They must be realistic with respect to available technologies, a matter of good technical judgment. If the services ask the impossible, or set requirements that drive up costs to the point that adequate numbers of a system cannot be produced (arguably the case for the B-2), the nation will not end up with the systems and equipment it needs. Second, requirements must fit the tasks that civilian officials are likely to assign to the military, whether or not service leaders welcome those tasks (e.g., peacekeeping). In both senses, "appropriate" technology poses a core challenge for acquisition and national security policy more broadly.

CHAPTER 8

COMPLEX SYSTEMS AND THE REVOLUTION IN MILITARY AFFAIRS

How much more complex than [chess] is the game of war, which occurs under certain limits of time, and where it is not one will that manipulates lifeless objects, but everything results from innumerable conflicts of various wills!

—Leo Tolstoy, *War and Peace*[1]

System performance depends on the way the pieces or subsystems fit together and function together. In "complex" systems the interactions may be subtle and surprising, and sometimes lead to unexpected failures. This chapter begins by exploring the reasons, then goes on to discus computer software and its ever-expanding place in military equipment. The second half of the chapter considers the Pentagon's much-touted Revolution in Military Affairs (RMA), which postulates a transformation of warfighting practices as a result of ongoing advances in precision weapons, sensors, computers, and communications—and the software that binds all these together. By comparison with the RMA, the most prominent military technologies of the Cold War—awesomely powerful nuclear weapons and the long-range bombers and intercontinental missiles that would have delivered them—seem brutally simple.

As illustrated by the accordion-like behavior of cars on a freeway, people are an integral part of many complex systems. (This example also shows that such systems need not have fixed identity: vehicle traffic is always in flux; so is the system of interconnected computers that makes up the Internet.) They add an element of unpredictability and at the same time resiliency or robustness, because people can often adjust to unanticipated events better than "hard-wired" or software-dependent system components. Because people are also sources of unanticipated events, operator "mistakes" are to be expected. The system must accommodate their actions—e.g., under the stress of combat—through warnings, corrections, or other built-in "fail-safe" mechanisms. If the system does not do so, the failure should be ascribed to design error rather than operator error.

Computer software adds new layers of complexity to military systems and the RMA promises to take these to new and unprecedented levels. RMA proponents sometimes compare the "system of systems" they envision to innovations in information technology (IT) and commercial computer and information systems. That is misleading. Commercial IT provides few guideposts and in fact suggests how difficult the RMA "vision" will be to achieve. The computerized information systems that have transformed business practices in many parts of the economy function best in stable, predictable settings, such as those encountered in financial services or retailing. Warfighting environments are anything but standardized and stable; sowing chaos in the ranks of the enemy is a primary objective.

The RMA also raises questions of doctrine, training, and command and control of a sort the services have faced and failed to surmount in the past. Without accepted doctrine, cemented by joint training, something more like confusion than coordination can easily emerge. Yet how can generic rules be developed? In the absence of warfighting experience? When tomorrow's enemy is unknown? If adversaries can fade into the background, vanishing among civilians and fighting in rag-tag groups as in Iraq, declining to present targets for 500-pound warheads guided by the RMA's system-of-systems? One of the lessons of past episodes of military innovation, such as strategic bombing, is that extrapolation far beyond actual warfighting experience is a poor basis for critical national security decisions.

System-Level Performance: The Parts versus the Whole

Complex technical systems are those that can usefully be decomposed, in actuality or in the abstract, into interacting subsystems. The connections or coupling between subsystems take on many forms. For instance, an airplane's wing and tail surfaces are coupled not only by the intervening structure but also by the air: the wake of the nose and wing affects flow over the tail surfaces and hence the plane's flight characteristics. The greater the complexity (crudely indicated by measures such as the number of nominally independent subsystems) the greater the likelihood of unexpected departure from the behavior predicted by mathematical (or physical) models. The B-1B's electronic countermeasures suite, consisting of more than 100 "black boxes" packed with electronics, functioned acceptably when tested in the laboratory. Crammed together inside the bomber, they interfered with one another—a problem that cost hundreds of millions of dollars to resolve. This is one example of a general point: complex systems, axiomatically, never perform as expected. If they did, we would probably decide that they were not in fact very "complex."

Military systems embody levels of complexity, extending to organizational complexity, with few parallels in the civilian economy. Air strikes during the 1991 Persian Gulf War required the orchestration of multiple waves of planes flown from different starting points by coalition pilots who had never trained together:

> One need only consider the immensely difficult balancing act of getting 400 coalition fighters airborne and marshalled at night in radio silence, refueled (often several times), and working under tight time lines without a missed tanker connection, let alone a midair collision or other catastrophic accident, to appreciate how aircrew skill and the ability to adapt under stress were critically important to the air campaign's outcome.[2]

While related technical and organizational complexities exist in hospitals, chemical plants, and the electric power grid, and complexity is hardly the only source of technology-related difficulties in military equipment, as we saw for the M16 rifle, defense systems exhibit many of the difficulties posed by technological complication in extreme form. System designers, for example, must anticipate contingencies such as battle damage having few if any counterparts in commercial systems.

The central problem in the design of any very complex system is to configure large numbers of components and subsystems, some of which may be poorly understood analytically, so as to yield the desired level of performance. There may be a large number of alternative arrangements and no good way to choose among them until a system has been built and tested. At that point, subsystems that worked work well in isolation may be found to function poorly when combined, as in the B-1B countermeasures example. The Army's Divisional Air Defense antiaircraft gun (DIVAD, mentioned in the preceding chapter) provides a copybook instance. Although designed from the beginning around available and proven components and subsystems, DIVAD prototypes never demonstrated even marginally promising performance.

As such examples suggest, system-level performance does not correlate closely with rates of technical advance in components and subsystems, such as integrated circuits (ICs). Complex systems can only be assessed by considering the system as a whole. That poses its own difficulties, both analytical and experimental. As we will see, computer software programs of any size cannot be tested in all possible states and operating conditions. Aircraft designers face a different sort of problem. They can call on well-understood mathematical models for predicting, say, air flow over a wing and the resulting forces of lift and drag. Once the wing is attached to a fuselage, the analysis becomes far more difficult. Some such problems can be handled by applying greater computing power. Others, like the wing drop encountered by the F/A-18E/F, cannot. The usual reason is limited physical understanding, so that the governing phenomena cannot be accurately represented in mathematical form. When this is the case, more powerful computers do not help much.

The more complex the system the greater the likelihood of deviations from anticipated behavior. The piping systems for nuclear powerplants include around 40,000 valves, 10 times the number in conventional powerplants. Lack of experience with these larger, more complex piping systems was a major reason for difficulties with nuclear power including higher than expected costs for construction and operation encountered during the 1960s and 1970s. Human operators add further layers of complexity. People make mistakes. The design of the system should take that into account. Everyday encounters with technology suggest how frequently it does not: we all struggle with nonintuitive computer software, sometimes even to turn on the intended burner on a kitchen stove—trivial tasks compared with flying a helicopter or interpreting blips on a radar screen.

When the U.S. Navy missile cruiser Vincennes shot down an Iranian Airbus on a routine scheduled flight in 1988, the chain of events began with misinterpretation by the Vincennes' officers and crew of information from their on-board Aegis fire-control system. No one realized they had mistaken the radar track of a passenger plane for an attacking F-14 (one of those the United States had sold to the Shah before the Iranian revolution). The Vincennes fired two missiles, killing nearly 300 passengers and crew aboard the Airbus. In the aftermath, Admiral Eugene La Roque called Aegis too elaborate and confusing:

> We have scientists and engineers capable of devising complicated equipment without any thought of how it will be integrated into a combat situation . . . . These machines produce too much information, and don't sort the important from the unimportant. There's a disconnection between technical effort and combat use.[3]

Others blamed organizational failure. In a penetrating dissection of actions aboard the Vincennes, Gene Rochlin ascribes the string of errors to a "false cognitive map" in the minds of the officers authorized to fire the missiles. After interpreting the initial Aegis information in terms of a "script" fitting a hostile attack, they failed to perceive the subsequent flow of information as inconsistent with such a scenario.[4]

Neither engineering tests and trials nor earlier operating experience foreshadowed the particular sequence of events that led the officers and crew of the Vincennes to misinterpret the information fed them by the Aegis radar. Among the lessons: complex systems must be tested by ordinary military personnel as well as technicians and engineers who know the equipment and its foibles, in order "to find out what happens when the young sailor . . . pushes the wrong button at the wrong time."[5] Before the disaster, other U.S. naval personnel had taken to calling the Vincennes "Robo-Cruiser" (alluding to the movie "Robocop") because of her aggressiveness.[6] Yet there is no indication that the behavior of the ship's crew was outside the bounds that those responsible for the design of the Aegis, its installation on ships of this class, and the operating procedures prescribed by the Navy might reasonably have anticipated.

Too often, designers overlook or underestimate variability in human performance. They blame operators for technical problems the design group should have foreseen and pass along the burden of minimizing errors and maximizing system effectiveness to those charged with instruction, training, and discipline. The General Accounting Office once estimated that at least half of all failures in military systems could be traced to "operator errors."[7] GAO's choice of words points to the fallacy. Operators *predictably* err. Systems should be designed to accommodate those errors, compensating in ways that will minimize their consequences. Absent compelling evidence that the designers have acted to reduce the probability and consequences of errors to the lowest practicable levels, operator errors should properly be considered design errors.[8]

The U.S. Navy's peacetime record in operating aircraft carriers and nuclear submarines shows that, given extensive training and practice, systems of great complexity can achieve consistently high levels of safety and reliability. Carrier crews practice flight operations constantly in order to maintain and refine organizational routines that depend on large numbers of independent checks and redundancies to reduce potential dangers. But these and other examples of error-avoidance do nothing to remove the prior obligation of designers to reduce possibilities for error and accident to the minimum.

Why does "human error" continue to be blamed for what should be considered design error? In part because analysis and design of complex systems calls for modes of thinking that are foreign to many engineers and scientists. The entire thrust of the physical sciences, and by extension the engineering sciences, has been to break down complex systems into simple "chunks" (components, subsystems) that can be modeled as if independent of one another: that is the reason complex systems have been conceptualized and discussed as above, in terms of coupled subsystems. The reductionist approach, powerful as it is, has two major drawbacks. First, for reasons already noted, the systems built up piece by piece do not always function as expected. Engineers and scientists are trained to anticipate and avoid these difficulties, at least when their origins lie in physical phenomena. But, second, human behavior is not deterministic and cannot be accurately represented statistically except under simple conditions. Thus human variability fits poorly or not at all with the usual reductionist methods for dealing with physical phenomena. The limitation is especially serious for predicting behavior under the extreme circumstances of warfighting, as exemplified by the imagined attack on the Vincennes.

Software

The B-2, designed in the 1980s, carries 200 digital processors; a replacement designed today might have 2,000. Pilots will fly the F-35 Joint Strike Fighter with the aid of 22 million lines of code. Software requirements for the Army's Future Combat System have risen to nearly 65 million lines of code. nearly doubling the initial estimate. If computers and software are everywhere in military systems, so are software errors. During the 1991 Gulf War, many of the bombs dropped by B-52s missed their targets by up to half-a-mile because of defective software. F-22s on their initial deployment to Asia in early 2007 had to stop in Hawaii for "repairs" before proceeding to Okinawa. Their navigational computers crashed because software developers had neglected to provide for crossing the International Date Line.

The Department of Defense (DOD) estimates that software now accounts for some 40 percent of all research, development, test, and evaluation (RDT&E) spending.[9] Many billions of dollars from other accounts pay for software maintenance—debugging and upgrades to existing code. For military and for many types of commercial systems, life-cycle costs for maintenance exceed those for the original development of code.[10]

Performance measures for computers and their hardware components have advanced at extraordinary rates for decades, as illustrated by Moore's

Law. Real prices for computers (i.e., as adjusted for performance gains) exhibit reductions by a factor of perhaps 3,000 over the past half-century, a rate of change unprecedented in economic history.[11] There is little sign that cost declines and performance increases will slow any time soon. But declining hardware prices do not translate directly into gains in system capability. Digital hardware cannot function without software, and measures of software advance do not show remotely comparable increases. With slow improvement in performance metrics for software and rapid gains for hardware, software now accounts for the majority of costs in many digital systems.[12]

Software poses unique technical problems. These stem from lack of proven methods for designing and testing programs of any size. Programming remains labor-intensive, the division of labor crude, management methods undeveloped. Programmer productivity, most commonly measured by lines of accepted code, differs among individuals by factors of 10, 20, even 100 or more. There are few accepted rules or norms for program design and testing. No matter how extensive and elaborate, tests cannot find all the errors or bugs in programs of much size.

In the early years of computing, when programs were written in machine or assembly languages (rather than higher-level languages such as Cobol or DOD's attempt at a standard, Ada), code generation was an onerous but manageable task. Software programs were compact by today's standards, in part because computer memory was too limited and costly to permit storage of large programs. Engineers and scientists wrote their own code for controlling laboratory equipment or steering guided missiles, analyzing laboratory data or performing numerical calculations earlier carried out manually. Because most of the applications were well understood, programs could be checked and verified against known results.[13] Then, starting around 1980, declining hardware costs led to progressive relaxation of the constraints on both memory costs and processing speed that had kept most programs, military and commercial, reasonably small. Over a dozen years from the mid-1980s, Microsoft Word grew from 270,000 lines of code to 2.7 million. As programs grew, they became much more time consuming and expensive to design, develop, and test. The frequency of bugs, or undetected errors, rose, not because of any increase in the frequency with which programmers made errors, but because errors are more difficult to detect in larger programs.

During the conceptual stage, software generation resembles other engineering design tasks, with a near-infinite number of alternatives from which to choose. But because software is itself an abstract symbolic construct and cannot be visualized or modeled like other engineered artifacts, the usual approaches to selection of a conceptual design and its development and refinement, such as mathematical modeling and simulation, cannot be applied. The only way to determine if a program will work is to write the code and run it. For the system as a whole, that means running *all* the code, under all possible contingencies—which is, however, impossible. Depending on the number of loops, branches, and subroutines, the number of possible end-to-end execution paths in even a relatively modest program may exceed 10^{20}. There is no conceivable way to check them all. Tests can only be based on selected operational scenarios.[14]

It may be equally impossible to check all possible operating modes of a physical system, or even all representative operational conditions. In such cases, engineers fall back on physical laws to model the system and predict its behavior (if imperfectly). Software developers have no such basis for prediction. They can only follow the often-proprietary testing and verification procedures imposed by their employer, plan and carry through internal "alpha" tests, followed by "beta" tests in the hands of nominally representative users, and then wait for field service experience to reveal further bugs. While most of these will have trivial consequences (and some may be invisible to users), a few may be serious (e.g., causing the entire system to crash).

Given the limitations of testing, software developers have no choice but to release programs that contain undetected errors. Their frequency commonly falls in the range of 1–10 per 1000 lines of code, which means that big programs contain thousands of bugs (over 60,000 when shipments of Windows 2000, with 35 million lines of code, began).[15] Some will not be triggered until others are found and removed. The corrective "patches" installed for this purpose sometimes introduce new errors.

After many years of use, large software programs may come to be regarded as "scrubbed" and reliable. Even so, latent defects are likely to remain, especially in blocks of code the system rarely exercises. This may be the bulk of the code, since software for a complex system must be capable of handling all possible operating contingencies, not just those routinely encountered, and there may be a great many of these. Latent defects sometimes trigger quite unexpected failures, such as occurred in August 2003 when a previously unknown bug in software controlling the electric power grid in the northeastern United States and Canada contributed to a chain of events that triggered a major blackout.

Software differs in a second respect from other technical activities that, like design more generally, have little in the way of a scientific basis. Designers regardless of technical field draw on extensive repertoires of heuristics and rules-of-thumb, some of them personal and others widely known and applied. Software has its own set of conventional practices, but they are immature compared, for instance, to those for the design of aircraft, or for that matter IT hardware. In the early years of computing, end-users trained in other disciplines, generally engineering and science, wrote most of the programs. As IT applications spread through the economy, demand for programming skills grew faster than supply. Since the establishment of academic computer science and engineering programs in the mid-1960s, they have graduated a cumulative total of perhaps half as many people as the 3 million or so currently employed in software occupations in the United States. Many software jobs, in other words, continue to be filled by people trained in other fields who have moved laterally through on-the-job learning and self-education. That is part of the reason a stable set of practices has been slow to develop, win acceptance, become codified, and diffuse.

In other engineering work, the need for "artful" choice narrows once preliminary design decisions have been made. The primary tasks then shift to analysis and refinement based on mathematical models and methods, followed by developmental testing. No such shift occurs in software projects, which remain art rather than (engineering) science to the end. Computer-assisted software engineering tools and methods have not as yet met expectations. Lacking metrics and tools, management of large-scale software projects remains primitive: "[DOD software] programs lacked well thought-out, disciplined program management and/or software development processes. Meaningful cost, schedule, and requirements baselines were lacking, ..." At the same time, DOD and its contractors seemingly do no worse than commercial developers: "[S]uccess and failure rate of DOD and commercial systems appears to be equivalent. . . . Studies reveal appalling performance in both environments."[16]

U.S. military and intelligence agencies operate a great many IT and command, control, communications, and intelligence (C^3I) systems. (The C^3I label has often been stretched or embellished to highlight favored themes or make sure no one feels left out; DOD has recently preferred C^4ISR, for Command, Control, Communication, Computers, Intelligence, Surveillance, and Reconnaissance.) These systems implement functions ranging from missile guidance and encrypted worldwide voice and data links to supply-chain logistics and maps for foot soldiers. They operate on networks that defy comprehension: diagrams of the communications links converging on an Aegis cruiser look like the proverbial rat's nest. Many of DOD's IT applications, finally, are old, running on long-obsolete hardware and the software equivalent of ancient languages. (Many commercial IT systems also run old software—so-called legacy programs—but business firms have had an easier time keeping their hardware, which does not have to meet specialized military standards and specifications, up-to-date.)

Although there are no good estimates, DOD's total software expenditures are very large and growing quite rapidly.[17] The Pentagon has little choice but to spend whatever it takes to get reasonable performance from software-intensive systems, old and new, else much of the military's equipment would not work at all. Businesses in the civilian economy share some of these problems, such as legacy systems, but not others, for reasons explored in the rest of the chapter.

Digital Dominance? Information Technology and the Revolution in Military Affairs

Computers and information technology, with their associated software, provide building blocks for the Revolution in Military Affairs, enabling "rapidly growing potential to detect, identify, and track far greater numbers of targets, over a larger area, for a longer time than ever before, and to order and move this information much more quickly and effectively than ever before."[18] RMA proponents believe that ongoing advances in C^3I, countermeasures, and precision engagement are coming together in ways that fundamentally transform warfighting. Because the United States is far ahead of other nations in these technologies, they believe the RMA will usher in a new era of "battlespace dominance" that will make it possible for the nation's military to win the battle and win the war without resort to the attritional strategies of the past. With the U.S. military's new intelligence and information systems guiding stealthy precision weapons fired from stealthy platforms, fewer warheads and fewer delivery platforms will kill more of the enemy and destroy more of his "assets." The RMA will reduce the significance of what military strategists call mass (numbers of men and tanks, planes and warships). Information will become the primary factor in determining battlespace outcomes. Thus when the United States invaded Iraq in 2003 it did not repeat the B-52 carpet bombing that had so stunned and demoralized Saddam Hussein's forces in 1991. At the same time, one of the potential fallacies of RMA scenarios was exposed for all to see, as the United States sent a force into Iraq capable of capturing and holding military targets but incapable of maintaining order otherwise.

Automated information processing, ever-faster communications, and increases in the accuracy and lethality of precision weapons do not by themselves make a revolution. Thousands of military IT and C^3I systems are already in use and technical improvements in precision weapons have been ongoing for decades. What makes a revolution, in the eyes of believers, is not so much advances in "enabling" technologies as their "seamless" linkage in a "system of systems" with "platform-centric" tactics giving way to "network-centric" warfare. The RMA, in other words, will be marked by a shift away from decentralized local decision-making by those in command of ships and army units to an integrated ("all arms") force consisting of many platforms fighting as a single organism under the control, figuratively speaking, of a single brain and nervous system. That brain and nervous system will be the core of the system of systems. All this will finally dispel the "fog of war," yielding a quantum jump in warfighting capability.[19]

Technological revolutions are rarely evident in the moment. No one can be certain, early in the twenty-first century, whether an RMA has arrived or may never quite arrive. After all, ostensibly revolutionary technologies have often been oversold, while truly revolutionary innovations such as the Internet sometimes materialize virtually unannounced. The remainder of this chapter considers RMA scenarios from two perspectives: comparisons with commercial IT and implications for command, control, and doctrine. RMA enthusiasts tend to overplay the first and overlook the second.

Business investments in IT hardware and software provide the closest parallels in the civilian economy to the RMA. These began in the 1960s when large firms began to automate payrolls, accounting, and inventory management. Other successes followed, including new products and processes in retail banking and finance (ATMs [automatic teller machines], near-instantaneous credit/debit card transactions) and the electronic reservation systems built by several major airlines. Still, the overall record is mixed and it is almost as easy to list disappointments: the paperless office and workerless factory; clinical (as opposed to administrative) IT in health care; once-heralded applications of artificial intelligence such as language translation.

Even the more sober RMA discussions rarely acknowledge past setbacks in military or civilian information systems.[20] They ignore almost entirely the clash between traditions of service independence, visible in everything from the conduct of joint operations to selection of equipment, and the "synergy" on which the RMA vision depends.[21] Proponents are more likely to celebrate an IT-based revolution in business affairs, suggesting some sort of parallel with the RMA. That is deceptive. Analogies with a business firm's processes and procedures are partial and imperfect. Most obviously, business organizations operate in relatively stable environments. They do much the same thing every day. Banks use IT to automate ATM deposits and withdrawals and write home mortgages and small business loans according to standard templates. These are variations on a theme, as are hospital billings and insurance reimbursements (administrative IT in health care has been at least a modest success, notwithstanding the complaints generated by poorly designed and duplicative forms and high error rates in data entry, which are mostly attributable to weak competitive forces in this sector of the economy). Airlines seek to maximize revenues through "yield management," making use of computer algorithms that juggle tens of thousands of fares more-or-less continually to fill as many seats on as many flights as possible (for the airline, an empty seat represents revenue lost forever). Their schedules change only occasionally and with advance notice.

Militaries, by contrast, aim to destabilize the environments in which they operate. Violence is not only an end in itself but a means of disrupting the opponent's ability to wage war. Whereas business transactions depend on shared understandings and common expectations, militaries sow mistrust and seek to degrade social cohesion through misinformation and propaganda. They go about their business by severing communication channels and transport links, depriving enemy armies and civilian populations of fuel and food.

For business firms, disruptions are anomalies. Their IT systems are designed accordingly. Because militaries are in the business of creating anomalies, their systems must work in disorderly and unpredictable settings, under physical and electronic attack from opposing "information warriors," including spoofing, jamming, and attempts to inject false information. Despite these and other differences, RMA advocates have sought to buttress their arguments with appeals to commercial IT such as Wal-Mart's computerized inventory control and ordering systems.[22] These have relevance mostly for logistics, a critical military function but one that is peripheral at best to the RMA.

Standardized business processes fit the strengths of IT. High volumes, low costs, and minimization of errors contribute to productivity and profits (much like assembly lines in mass manufacturing). On the other side of the benefit-cost relationship, firms have had to invest large sums year after year in hardware, software, and organizational change, including employee training, simply to keep pace with rivals. "Reengineering" during the 1990s proved unexpectedly costly for firms that found themselves struggling to restructure jobs requiring mastery of 100 or more software programs (e.g., for customer service representatives in some banks and insurance companies).

Even when operations are standardized and stable, firms sometimes fail to benefit from IT because the system ends up supplementing, supporting, or automating processes that do not function very well to begin with (a common pitfall in manufacturing automation too). Even if the intent is simply to strip out labor costs, the old process must be carefully defined and documented. Because business processes often grow up piecemeal over many years, no one in the firm may grasp just how they actually work.[23] Instead, the knowledge is held collectively (and is heavily tacit). This means IT design must start with detective work to understand the old system, which typically reveals ambiguities, contradictions, and inconsistencies that must first be resolved. In the end, not a few organizations have ended up simply replacing old and obsolete IT systems with "new technology" that yields no more than marginal improvements in operating performance (sometimes the new system is scrapped). The U.S. military has had its share of similar experiences, for similar reasons.

Legacy systems complicate the problem. As long ago as the 1960s, banks began to install software for automating high-volume products such as auto loans. As deregulation permitted entry into new lines of business, banks added software for money-market accounts, credit cards, home-equity loans, mutual funds, and so on. As successive mergers brought incompatible IT systems under the same corporate umbrella, some banks found themselves with several dozen software packages for a single application. The U.S. military likewise finds itself with overlap and redundancy among legacy programs written in different places at different times for similar, related, or identical applications, including more than 600 information systems just for logistics.[24]

Legacy systems raise the costs and increase the complication of RMA-style networking. The intent of "seamless" integration is to sweep away such impediments. But the task is enormous and has almost certainly been understated, just as many businesses and the consulting firms that advised them underestimated the costs, effort, and risks of business process reengineering during the 1990s.

Strategic applications of IT, intended to create lasting business advantages through innovations in products (or processes—product-process distinctions blur and break down for the service products that dominate output in postindustrial economies such as that of the United States) provide the closest, if still quite distant, parallels to RMA scenarios. For new products, just what customers want or will accept is initially uncertain. As a result, the IT system may have to function in an ill-defined setting until preferences emerge, to be explored and incorporated through software upgrades. Not surprisingly, IT performance has been generally weaker and over-promising more common compared to projects in which the objective has been cost reduction through automation of standardized products or processes.[25]

A closer look at innovations in supply chain management, as in retailing, indicates how easy it can be to overlook the limitations of IT. While the casual observer might assume that point-of-sale scanning enables retail stores to track stock levels accurately, in-store audits commonly reveal discrepancies in the range of 10–15 percent, with computerized inventory control systems reporting more items on the shelves and in the stock rooms than can be located. Human frailty, in the form of "shrinkage" (shoplifting and pilferage by employees), accounts for much of the shortfall. Errors also occur in logging goods into the store and at check-out. Discrepancies likewise afflict DOD's inventory control systems (which number about 200, a subset of the 600-plus logistical systems mentioned above). If the computer shows, say, 78 laser-guided bombs at a forward air base, are they really there? Or should mission planners ask for a count before finalizing tomorrow's air tasking orders (ATOs)? And if several shoulder-fired antiaircraft missiles cannot be located, might they have been stolen by locally hired workers and sold to irregular combatants who will be lurking outside the base perimeter when the planes take off?

What the system says, in other words—or seems to say, as in the Vincennes case—may not be true. Perhaps the information will be good enough for decision-making, perhaps not. Intelligence and counterintelligence, collection of information and dissemination of disinformation, are staples of military practice. These activities will always raise questions that technology, unaided, cannot answer. Is that yellow school bus full of kids or men with guns? If an unmanned aerial vehicle (UAV) detected enemy soldiers entering a warehouse two hours ago, are they still there? Or might they have left, to be replaced by refugees seeking shelter from the rain that has hindered surveillance in the meantime?

Critical RMA tasks depend heavily on interpretation of video, radar, and infrared images, "data fusion" to combine inputs from many different sensors, and processing of huge volumes of signals intelligence (e.g., intercepted communications). Many of the underlying hardware technologies are advancing rapidly: U.S. forces will at some point have sensors that can see unhindered by rain, indeed will be able to penetrate the walls of buildings and the roofs of school buses. On the other hand, the pace of change in technologies such as optoelectronics is fundamentally mismatched with DOD acquisition, which is glacial by contrast. It is not at all clear how DOD and the services will be able to avoid rapid obsolescence in RMA-related systems without spending ever-increasing sums, year after year. Businesses buy most of their IT hardware and much of their software off the shelf. DOD, heavily dependent on specially developed hardware and software (e.g., for the Army's Future Combat System), cannot. Businesses pay a high price to keep up; the price to DOD is higher.

System-level capabilities keep improving, of course. During the 1991 Gulf War, U.S. commanders and their Coalition partners prepared ATOs on a 72-hour time cycle. With "dynamic tasking" during the 2003 invasion of Iraq, planes could take off without final orders; they were sent their mission assignments on approach to target areas. While orders were mostly based on information that was several days old, planners could direct or redirect strikes to high-priority targets within an hour or so of detection by remote sensing or ground observers.

The full-blown RMA vision combines the assembly and interpretation of real-time sensory information from platforms such as UAVs and satellites, able to track all forces, friend and foe, in a battlespace extending from sea floor to earth orbit, with strikes—e.g., by weapons–carrying UAVs, by hypersonic cruise missiles that can be redirected in flight to moving targets before the latter can accomplish their own missions—vectored in near-real time. (Verification of ambiguous information will continue to impose time lags.) The "sensor-to-shooter" links of the old systems will be replaced by a command and control hierarchy that actively manages all friendly platforms in a three-dimensional battlespace. This is not a new idea: "The United States has been striving for a single integrated air picture since 1969 when the Tactical Air Control System/Tactical Air Defense System was launched."[26] But the vision has not been attained.

If and when it is, command authorities will be able to achieve an age-old objective: bringing their own greatest strengths asymmetrically to bear in concentrated strikes at an opponent's points of greatest weakness (perhaps fleeting). The U.S. military will disable its opponents with stand-off weapons while minimizing the exposure of friendly forces to the bloody business of battle. As the war begins, precision strikes will surgically eliminate whatever battlespace awareness the enemy may possess. Space-based weapons, for instance, might target an adversary's satellites while protecting those of the United States and its allies. The enemy will be blinded, his sensors snuffed out, struck dumb by the severing of his communications links, hamstrung by the evisceration of his computer-based C^3I systems. All exits barred by fast-moving mobile units guided by remote sensing platforms and the intact communications links of the United States, the enemy will be unable even to flee. With near-perfect battlespace awareness, an intelligible picture of events in near-real time, U.S. forces will deploy their precision weapons to near-perfect effect, decapitating the enemy's leadership, breaking through his defenses at points of their own choosing and isolating vulnerable units to be picked off at leisure. The United States will be like Muhammad Ali dancing circles around a blindfolded opponent.

(Or, as it might be phrased within the Pentagon, "The Air Force's precision engagement core competency will enable it to reliably apply selective force against specific targets simultaneously to achieve desired effects with minimal risk and collateral damage."[27])

More than 30 years ago, General William Westmoreland announced that

> On the battlefield of the future, enemy forces will be located, tracked and targeted almost instantaneously through the use of data links, computer assisted intelligence evaluation, and automated fire control.
> I see battlefields or combat areas that are under 24 hour real or near real time surveillance of all types. I see battlefields on which we can destroy anything we locate through instant communications and the almost instantaneous application of highly lethal firepower.[28]

U.S. forces could not accomplish this in the Persian Gulf in 1991 or Kosovo in 1999, any more than in Vietnam. In Afghanistan during early 2002,

> [A]n intensive prebattle reconnaissance effort focused every available surveillance and target acquisition system on a tiny, ten-kilometer by ten-kilometer battlefield. Yet fewer than 50 percent of all the al-Qaida positions ultimately identified on this battlefield were discovered prior to ground contact.[29]

As Operation Anaconda continued, fragmented command structures and communications networks meant that

> [U.S. forces] sometimes found themselves talking to the . . . fires officer over a thousand miles away in Masirah in their efforts to get an aircraft overhead to strike a target only a couple of thousand meters in front of them. The result . . . was that . . . troops were filling the radio nets with calls for close air support . . . . [Observers] could identify the mortar positions and machine guns firing at the infantry, but sometimes had to wait over an hour to arrange for an air strike on the target.[30]

Similar situations had been commonplace more than 50 years earlier during the Korean War, as recalled by Marine Corps General O.P. Smith:

> General [Earle E.] Partridge commanded the 5th Air Force, and he'd been raised on the Joint Operations Center business where everything had to be cleared through the Joint Operations Center [JOC] at Taegu.
> Finally we got Gen. Partridge to come up to our [Command Post], when we were somewhere up near Hongchong . . . . While we were talking, somebody had requested an air strike, and they were trying to raise Taegu to go through the JOC. It took 45 minutes to get the call through, and I told Gen.

Partridge, "You see what this involves?" But it wasn't changed, and what it meant was that our people up in the front lines, finding that it took so long to get any air, just . . . tried to get the artillery to do the job . . . .[31]

When might the full RMA vision be possible? Or will the RMA, like the paperless office, prove to be a misconceived goal? Relevant DOD experience includes the "long and turbulent histories of the JTIDS [Joint Tactical Information Distribution System] and MIDS [Multifunctional Information Distribution System] programs" and, more recently, cost overruns and performance shortfalls on software for the F-22.[32] Even purely technical steps have halting. According to GAO,

> [T]he systems involved in the sensor-to-shooter process do not operate effectively together. There are over 100 command, control, communications, intelligence, surveillance, and reconnaissance systems . . . . But these are separately owned and operated by each of the military services as well as other DOD and intelligence agencies. These separate systems have limited ability to interoperate . . . . [C]ommunications systems must be patched together to make this happen.[33]

The "system of systems," in other words, is far from "seamless" operation. That does not mean the goals will not be reached. But it will take a long time, and it is possible that implementation will turn out to be a receding objective, always in sight but never attained.

Although the RMA has been billed as banishing the fog of war, new technologies have potential for extending that fog. It was smoke from small arms and artillery that obscured the battles of Clausewitz's day, yet even then the significance of the "fog" lay not so much in any literal meaning as in the evocation of confusion, disorientation, lack of reliable information, rips in the communications net—anything and everything that might hinder decisive action at critical moments. Effective response to the flow of events depends on the ability of command authorities at several levels to grasp what is happening on the field of battle (or in the battlespace), separating signal from noise, identifying patterns before the enemy does so, distinguishing emerging threats from the opponent's feints and attempted deceptions, and otherwise operating inside the opponent's "decision loop." All of this must be accomplished with great speed: Arleigh Burke once told a junior naval officer, "The difference between a poor officer and a good one . . . was 'about ten seconds.'"[34]

RMA technologies promise to deliver ever-expanding volumes of data and information. Most of this will have to be analyzed before it can be put to use. Because too much partially digested information fogs the mind, command and control centers will have to sort and interpret it, "filtering" and "fusing" the incoming flow to create a coherent picture, fit for human consumption. Unknown uncertainties will characterize much incoming intelligence and surveillance data, which will have to be cross-checked and verified, scrutinized and interpreted—work that today takes hours, days, and sometimes weeks.

For command and control decisions in something approaching real time, information will also have to be presented in ways that mesh with doctrine. Doctrine itself is bound to change, although how this might happen seems not to have been explored in much depth. For centuries, military commanders have struggled to bring irresistible force to bear at a decisive point. Before electronic communications, they worked from fragmentary reports, frequently erroneous. Armies relied on scouts and messengers, ships on signal flags, at night on lanterns. Because no one could be certain of what was going on outside his own field of vision, command and control had to be decentralized: there was no other way. The commander and his staff planned the battle and issued orders, after which they often had little choice but to passively await word of the final outcome.

Telegraph, telephone, then wireless and voice radio made information transmittal quick and reliable, but did only a little to improve its quality. The RMA promises to provide accurate information quickly to decision-makers. As yet, planners have had little to say about how that information will be used at various levels in the hierarchy of command, or about how that hierarchy might change. The combat effectiveness of armies has always depended on the initiative and judgment of officers down to the most junior ranks (and on noncommissioned officers). Yet RMA scenarios are not likely to permit field officers and local commanders as much discretion as in the past if this could put at risk the large-scale coordination on which battlespace effectiveness depends. The RMA, in other words, implies a more rigid command hierarchy, for the Army and Marine Corps especially.

At the same time, if the entire battlespace is under surveillance in real or near-real time, field officers will expect access to any and all data and information with possible relevance to their local situation. After all, some seemingly obscure fragment, meaningless to headquarters, could have life-and-death consequences for the units under a field commander. Yet too many people with too much freedom of action risks the collapse of the larger operation in a blizzard of independent decisions, with the potential to put all units at risk. That has always been true. The difference is that an RMA promises both to expand access to data and information by everyone from low-level field officers to kibitzers in the Pentagon and to expand the spatial boundaries within which military action can, at least in principle, be "managed." Command authorities will seek coordination over wide areas to maximize striking power, win the battle, and win the war. They will have the capability and most likely the desire to make continual adjustments based on their sense of the overall situation, issuing a steady flow of guidance. Field officers, on the other hand, will want local information and the freedom to act on it. At least some of the time, the two imperatives will be in conflict.

One of the characteristics of large-scale, tightly coupled systems is that errors sometimes propagate unchecked with wide-ranging consequences, while in a loosely coupled system the same error might do only localized damage. In the past, decentralized command and control of ground troops reduced the likelihood of error propagation. Poor decisions were revealed, first of all at the scene, and corrective action could be taken. Decentralization increased the army's robustness (resistance to damage) and resilience (recovery capacity). Somewhat greater centralization of command and control is implicit in almost any RMA scenario and some robustness and resilience could be lost. Because rapid and unquestioning response will be critical for effective "battle management," officers who in the past would have had time to consider whether orders might be based on tardy, incomplete, or otherwise faulty information stand to lose some of their freedom of action. Doctrine and training will tell them that exercising too much discretion could put the larger operation at risk. At the same time, the unthinking responses demanded by systemic imperatives may increase the risks that the entire undertaking will come apart for unanticipated reasons (e.g., previously unknown software errors).

Performance in large-scale, tightly coupled systems normally improves with accumulated experience. That will no doubt be true of RMA-style battle management. Hardware and software will be debugged, technology and doctrine mutually adjusted, tensions between centralized command and local autonomy mediated if not resolved. Yet to reiterate another point that is central to the analysis in this book, actual warfighting, as opposed to games, exercises, and maneuvers, invariably brings surprise. During the 1991 Gulf War, lower-level officers found it easy to get information through informal channels. With secure telephones they could call a friend or acquaintance in Riyadh or even in Washington:

> The [secure telephone's] ease of use, hardly more difficult than an ordinary phone, and its wide distribution throughout the DOD had an almost subversive influence on war planning and execution. It gave junior officers and subordinate units the ability to circumvent the chain of command and inconvenient,

slow, or misunderstood practices at the cost of creating confusion. All command elements no longer precisely understood what information other elements had used as the basis for their actions.[35]

Some were tempted, when instructions from higher command were late in arriving, ambiguous, or seemed less than sensible, to act on their own. That sort of unanticipated behavior was a consequence of a modest increment in access to information, in nothing like real time. RMA scenarios promise far more dramatic shifts in patterns of information flow.

For more than a decade after the Korean War, the Air Force and the Army debated air support for ground troops. Mixed in with the posturing and turf battles were concrete questions that merited thinking through and codification: How many daily sorties could the Air Force guarantee? Did effective close air support require Air Force procurement of planes designed for that purpose? Where did the Army's helicopters fit in? Which service should have control of mission planning? Control of aircraft once over the battle zone? Did effective control depend on command authority? Or could control and command be split between Army and Air Force officers? Five decades later these issues have not been fully resolved. Will the appeal of the RMA vision finally accomplish that? Or will matters devolve into "the little league definition of jointness . . . 'everybody's on the team, and everybody plays.'"[36]

CHAPTER 9

INNOVATION AND LEARNING IN PEACE AND IN WAR

I . . . remember looking down . . . and counting the fitful yellow-orange flares I saw on the ground. At first . . . I did not understand them. Here were no cities burning. No haystack could make a fire visible in broad daylight 23,000 feet up. Then it came to me as it came to others—for I remember my headset crackling with the news—that these were B-17s blazing on the ground.

All across Germany, Holland, and Belgium the terrible landscape of burning planes unrolled beneath us. It seemed that we were littering Europe with our dead.
—Elmer Bendiner, *The Fall of Fortresses*[1]

The puzzle this book seeks to assemble is comprised of pieces including technological innovation, warfighting practices, and acquisition politics and policy. Earlier chapters alluded to the differences between military and commercial innovation. The parallels, though loose, are revealing, the contrasts more so.

Some radical innovations begin with "breakthrough" inventions such as the transistor. Others stem from "normal engineering" that initiates (or continues) a sequence of improvements guided by market signals. As we saw in chapter 7, this describes the evolution of the microprocessor. And radical innovations sometimes result from the coalescence of previously separate and independent technological streams, as illustrated by the Internet.

These phenomena have relatively few points of contact with innovation in military systems, which in the post–World War II period the U.S. Department of Defense (DOD) sought to manage deliberately. Like the Internet, many military systems embody multiple and more-or-less independent technological developments. Unlike the Internet, the absence of market signals and related forms of feedback from users vitiate efforts to "insert" continuous improvements into weapons systems. The Internet emerged in seemingly spontaneous fashion. What could be more remote from the calculated and constant political and bureaucratic maneuver of DOD acquisition?

The chapter opens with a brief analytical overview summarizing varieties of technological learning and their contributions to innovation. It then presents three case examples in enough detail to illustrate the analytical points concerning innovation discussed below and in the next chapter. The first two examples continue the story of strategic bombing and give a brief account of the demise of the battleship. The technologies available in the 1930s were far from adequate for strategic bombing. Largely through inattention at high policy levels, the United States slid into acceptance of an untested "theory" (one that in hindsight does not appear sufficiently plausible to merit the label of theory). Only after four decades and the equivalent of a technological revolution in the form of precision weapons did strategic bombing become a reasonably efficient instrument of military power. In the case of the battleship, long the centerpiece of naval fleets, advances in aviation rendered an existing technological system obsolescent.

The final example, civilian nuclear power, is very different. For comparative purposes, nuclear power is more revealing than such unequivocally positive spin-offs from defense as electronic digital computers. In the nuclear power case, Cold War geopolitics led the U.S. government to subsidize purchases of light water reactors based on designs originally developed for the Navy. In a sharp break with their traditional practices of conservative technical choices and investment decisions, utilities committed large sums before gaining experience with the operation or even the construction of nuclear powerplants. The government's subsidies and salesmanship short-circuited learning processes that in the civilian economy generally proceed at a more measured pace, distorting feedback signals that under normal circumstances guide and channel innovation.

Innovation, Military and Commercial

Military and commercial innovation differ in fundamental ways. Market-mediated feedback guides and steers innovation in the civilian economy. These mechanisms have few counterparts in defense, making it hard to recognize the wrong turns and mistakes that are inherent in all innovation. In the civilian economy, competition pushes firms to emulate or outstrip their rivals. The market tests these efforts continuously, ruthlessly exposing ineffective or unwanted innovations. Firms make ongoing adjustments in response to these signals; those that do so most effectively prosper. Powerful and reliable feedback into military innovation comes only during wars—and only wars fought against capable adversaries. Modes of competition other than war, such as international arms races and inter- and intraservice rivalries, do not create pressures comparable to those exerted by market forces in the civilian economy. Inter- and intraservice rivalries, in particular, are only weakly self-correcting because political and bureaucratic forces may, in the absence of the tests of warfighting, override considerations of military utility.

In both military and commercial spheres, innovation can be viewed in terms of learning. In the civilian economy, new knowledge and new ideas foster entrepreneurial ventures and commercialization. Sometimes these efforts succeed and sometimes they fail, at least initially. Sometimes new markets open where no apparent demand existed previously. Successful commercialization begets an ongoing search for improvements, as for the microprocessor and jet engine. Customer demand for greater performance (or a different combination of performance attributes) and lower costs induces further innovation. Many if not most of the improvements are incremental; they may be invisible except to those immersed in the evolving technology as suppliers or customers. Together, demand for and supply of incremental innovations guide and shape the unfolding process. Intel became the dominant supplier of microprocessors by catering to emerging markets more successfully than other entrants. That was not a consequence of Intel's early lead in commercialization, but of the firm's superior reading of what customers would buy. Boeing, similarly, did a better job than Douglas and Lockheed in the 1960s and 1970s of judging what airlines wanted in commercial transport planes.

In a market economy, three sets of participants contribute to the unfolding dynamic of innovation: firms that develop and supply the innovations themselves, their immediate customers, and end users. (Sometimes the customer is the end user; other times the customer incorporates the innovation into its own products or processes, as when automobile firms adopted microprocessors for powertrain control.) In three ways, intermediate and end users contribute to innovation.[2] First, they channel information (e.g., modes of failure, shortcomings and desired improvements) back to innovating firms, where it becomes an input to ongoing R&D and D&D. Second, users may devise applications that innovators did not anticipate. Third, they sometimes contribute directly to technical advance. Indeed, users on occasion diversify into design and production of goods or services they first encountered as customers, expanding the set of innovating firms; many software companies started this way.

Somewhat more abstractly, networks of feedback and feedforward loops link innovators with users. Based on feedback from users, innovating firms decide what improvements deserve highest priority. Through feedforward, they attempt to anticipate features users will value in the future. These mechanisms have few counterparts in military technological innovation. In the civilian economy, for example, independent owner-operators buy many over-the-highway trucks. They vote with their dollars for design features such as air suspensions. In the military, enlisted men and women drive trucks purchased by the Army. No one asks them what they want. The Air Force officers who pilot planes have more influence, but only after they have reached a high rank. Feedback from actual users, if it gets to system designers at all, is likely to be muffled. Negotiations between the services and their contractors take the place of market signals.

The evolution of jet and gas turbine engines illustrates the importance of user feedback. Following Pratt & Whitney's introduction of its JT4A engine in the late 1950s, technicians and engineers employed by airlines worked out maintenance methods that took less time, reducing flight delays, and saved money in heavy maintenance (by, for instance, enabling repairs without removing the engine from wing or airframe). Their engineering staffs meanwhile fed information back to Pratt & Whitney and its suppliers (e.g., on components that wore out quickly or failed prematurely and also on those that showed little wear at the end of required maintenance intervals). Cumulatively, the improvements were such that, over the JT4A's first six years in commercial service, average time between overhauls rose from under 1,000 hours to over 6,000 hours.[3] While the armed forces also channel information back to engine manufacturers and their suppliers, they accumulate flight hours slowly compared to the airlines. They also have much weaker incentives to reduce (or even track) operating costs, which affect airline profitability directly but for the military are just another line item in the budget.

Three basic forms of technological learning can be distinguished, illustrated below for jet engines and gas turbines:

- *Learning in R&D/D&D.* Over many years, engine manufacturers and their suppliers developed better surface coatings for protecting high-temperature components such as turbine blades from oxidation and erosion. The coatings extended service life and also permitted higher operating temperatures, which leads directly to increases in thrust and fuel efficiency.

- *Learning in production.* Also over many years, employees of engine manufacturers and their suppliers—production workers, technicians, engineers—found ways, mostly through experiential learning, to reduce labor inputs in component fabrication and assembly.

• *Learning by users.* As already described, user learning fed back into R&D and D&D.

These modes of learning are related but distinct. Coatings for engine components stem primarily from laboratory research. Opportunities for learning in production, on the other hand, cannot in general be foreseen in advance (else they would have been implemented by engineers and technicians during process design) and rarely have much to do with R&D.[4] Complicated assemblies such as gas turbines, for example, can be put together in dozens, hundreds, or even thousands of sequences, many of them not readily apparent. Only for the simplest assemblies is it obvious which of the possible sequences will be quickest and easiest. Until the past decade or two, the best assembly sequences could only be found by successive trials on the shop floor; today computer simulations speed the process.

These three types of learning are pervasive in the civilian economy. Incremental technological change is rapid and continuous; it happens all the time and almost everywhere. This year's personal computers and supercomputers come with faster chips and more memory. New cars feature "smarter" air bags, modified control algorithms for engines and transmissions, headlamps that throw more light on highway shoulders. The same three modes of learning also contribute to military technological innovation, but in attenuated form. The reasons include lengthy intervals between redesigns, which makes continuous improvement difficult, low output, which limits learning in production (and can lead to unlearning if, for example, procurements are interrupted), and sparse and sporadic feedback from warfighting, which constrains and sometimes distorts learning-by-using (table 7).

Because DOD spends so much money on R&D, learning through research goes on constantly. But opportunities to *incorporate* new knowledge come only at widely spaced intervals. Decades sometimes pass between platform designs. While new weapons and new electronics can be fitted to old platforms, space and compatibility constraints limit the resulting performance improvements. Then, when an all-new program finally begins, the services quite naturally try to cram in everything possible in the way of new technology. By far the most important limitation, however, stems from the intermittent nature of actual combat. There is no way to know the capabilities or predict the actions of a resourceful and determined foe in the absence of war. When war does come—and one purpose of military innovation is to deter war—it invariably brings surprises. Given the exigencies of combat and the press of events, there may be little opportunity to reflect upon and absorb the

Table 7 Technological Innovation, Commercial and Military

Commercial	Military
Product cycles in many industries short and getting shorter.	*Acquisition cycles* long, getting longer.
Economic *competition* creates incentives for innovation affecting both product attributes and manufacturing processes.	Market-based *competition* seldom significant; when present, tends to be weak relative to political and bureaucratic forces. Although design competitions may be practical for large procurements such as the F-35 Joint Strike Fighter, competitive bidding for production contracts is rarely feasible except for expendables consumed in large volume, such as munitions.
Adoption paced by market forces drives competitive emulation; when one firm introduces technological improvements, others feel compelled to follow.	*Adoption* of major innovations requires consistent and strong support by one of the military services, which must be politically constructed.
Learning in R&D/D&D, including engineering tests, is a major source of product/process innovation. Feedback from the field guides continuing improvement.	While *learning in R&D/D&D* generates new system concepts, years may pass before a new program permits reduction to practice. Engineering tests often lack realism because rigorous trials risk exposure of shortcomings with possible damage to political support.
Learning in production is a major source of reductions in cost and increases in quality.	Output volumes small and getting smaller, reducing opportunities for *learning in production*.
Learning by users contributes to functional improvements in product attributes, gains in operating efficiency, and, less commonly, to decreases in manufacturing costs and increases in quality .	In peacetime, *learning by users* is restricted to training, exercises, and maneuvers. These cannot fully explore a system's strengths and weaknesses and may fail to reveal how well technology and doctrine mesh. In wartime, the risks of combat elevate the costs of experiential learning and may hinder extraction and absorption of lessons (e.g., if lacking obvious and immediate relevance to the current conflict).

lessons or to try out technical modifications and doctrinal revisions. Because lessons learned cannot be validated until the next conflict, nothing like continuous improvement is possible.[5] Some lessons may be resisted by those with reputations and interests to protect. Others may turn out to be wrong and have to be unlearned in the next conflict, perhaps at high cost in lives. Decades of inaction may be punctuated by bursts of almost unimaginable violence, followed by years of pondering outcomes amidst recriminations and scapegoating. In the civilian economy, market mechanisms ordinarily reveal poor decisions relatively swiftly, keeping individual and organizational interests reasonably congruent. Forces for self-correction are weaker in military organizations, especially during peacetime but also in times of war.

Metrics exist to capture the three types of learning in at least some of their dimensions. Engineering parameters (e.g., thrust and efficiency for jet engines) track advances resulting from R&D/D&D. Accounting conventions underlie measures of learning in production. Some combination of engineering and accounting measures can represent many improvements due to learning by users. More generally, however, integrative measures of effectiveness do not exist. This is true for many commercial and essentially all military systems.[6] The reasons lie in the incommensurability of available indicators. In the civilian economy, market outcomes sort out the relative value, to take a simple example, of access time and storage capacity for computer memory. Nothing similar takes place in defense. Performance measures and their significance may become contested inside DOD or between the services and their contractors . At the same time, the questionable relevance of available metrics tends to be hidden from outsiders by the military's claims of professional knowledge and sometimes by secrecy. Thus former Air Force Chief of Staff Ronald Fogleman defended the F-22 by declaring that "In the 'black world,' the F-22 is a truly revolutionary airplane. On the surface, it looks conventional, like an F-15 with some stealth capabilities. But the combination of stealth, supercruise, and integrated avionics is a quantum jump."[7] General Fogleman (himself once an F-15 pilot) did not try to explain just how these capabilities come together to constitute a "quantum jump."

Strategic Bombing: The Long, Slow Road to Precision

Forty Years of Technological Inadequacy

Sometimes technology and doctrine are closely interdependent. In other cases, technical attributes do not matter much. During World War II,

[S]uccessful close air support . . . depended upon organization, doctrine, and training, rather than upon the existence of specialized weapons systems.

"Strategic bombardment" in daylight required two essential technical ingredients: the long-range, four-engine bomber and the high-performance escort fighter to accompany it. Air forces that failed to procure the appropriate aircraft types . . . found themselves in an unenviable predicament . . . . Close air support did not depend upon the same essential "technical fix" . . . .[8]

For ground support, in other words, effective employment of whatever aircraft were available outweighed the performance characteristics of those aircraft. Long-range bombing, conversely, had no hope of success without suitable aircraft, not only bombers but escorts capable of fending off enemy interceptors.

At first, the U.S. Army Air Forces had the bombers but not the fighters, for reasons explained earlier (chapter 6). Even if long-range escorts had been available, however, prevailing doctrine—daylight precision bombing of "strategic" targets—could not have succeeded. There were four main reasons: (1) U.S. planners had incomplete and inaccurate information on potential targets (poorer still for Japan than Germany); (2) they overestimated the effects of bombing on Germany's (and Japan's) war-making capacity; (3) the bombers could not reliably locate targets selected for them; and (4) when bombers did succeed in finding their assigned targets, they could not hit them. The first two difficulties were basically matters of economic intelligence, the latter two of technological capability. Any one of the four would have been a severe handicap. Taken together, they made it nearly impossible to do serious damage to targets of strategic value until late in the war, when German defenses, and especially the Luftwaffe's ability to put interceptors into the air in numbers, had been so weakened that Allied bombers could mount massive saturation attacks. These finally began to degrade Germany's warmaking capacity.[9]

There are few better illustrations of the myopia that sometimes afflicts military organizations than the Army's long-lasting embrace of precision bombing, which in practice was nothing of the sort. We have already seen that this was the preferred mission of the Air Corps, for which it had a singular virtue: only the remotest connection with the ground army. Once free to embrace their preferred mission, aviators shunned tests and trials, barely pausing to consider the accuracy with which bombs could be delivered under ideal much less warfighting conditions—i.e., searching for hard-to-find targets under hostile fire, targets that might be unrecognizable from high altitude even in clear weather.

Strategic bombing advocates preferred to avoid questions of economic intelligence. World War II planners lacked sufficient knowledge of the structure and functioning of German industry to choose objectives sensibly.

According to a hastily prepared 1941 study, there were 154 key targets, no less and no more, the destruction of which would cripple the Nazi war machine.[10] This was wildly off, yet in the absence of better information continued to influence target selection until the end of the war. (In practice, weather conditions often dictated those targets, if any, that could be attacked on a given day.) Planners had even less insight into the economy of Japan. While U.S. forces managed to strangle Japan's wartime industrial output, they accomplished this not through the destruction of targets on land, although much destruction was wrought, but by sinking ships at sea. As U.S. forces in the Pacific grew stronger and Japanese defenses weaker, tankers and cargo vessels carrying raw materials and oil to the home islands went to the bottom faster than they could be replaced.

When the time came to put strategic bombing to the test, navigation to target areas posed great difficulties.[11] Heavy losses in 1939, when the Luftwaffe's interceptors shot down unescorted bombers right and left, had persuaded the British to fly only at night.[12] The Americans were determined to persist with daylight bombing.[13] Once over target areas, if clouds or fog obscured the ground or the pilot tried to dodge hostile fighters or flak (more of a danger to B-24s than higher-flying B-17s), the best of bombsights, including the top-secret Nordens, proved close to useless, since anything approaching accuracy required flying a straight and level course with crosshairs fixed on the aim point; even then, the jet stream, over Japan especially, often swept the bomb-load off-target.

Weeks and sometimes months passed without clear skies over Northern Europe for visual bombing. Provided the ceiling in England was high enough to permit the bombers to take off and assemble, attacks could be conducted through cloud cover once radar bombsights became available, although these were good for little more than distinguishing city from countryside. When warheads did explode on targets of military or economic significance, they caused less damage than expected. Repairs were rapid. Germany dispersed its war industries. Civilian morale did not suffer, but stiffened, as it had among Londoners during the 1940–1941 Blitz. The Allies never seemed to realize that industrial facilities had to be struck again and again to hinder rebuilding and were slow to understand that they were doing relatively little damage, in part because their intelligence services produced greatly exaggerated estimates of declines in Germany's war production resulting from the bombing.[14] As realization sunk in of how little they were accomplishing, the Americans, like the British before them, grew less selective, portraying the flattening and torching of German cities as "strategic" by virtue of killing or injuring war workers and destroying their homes. A 1944 directive made explicit that almost any built-up area was a permissible target: "It has been determined that towns and cities large enough to produce an identifiable return on the H2X [radar] scope generally contain a large proportion of . . . military objectives."[15] This was a fig leaf.

By 1944, range-extending drop tanks became available for P-47 and P-51 fighters, enabling them to escort B-17s and B-24s deep into Germany. Radio and radar navigation and targeting aids, many based on British developments, helped improve accuracy. Although German antiaircraft fire still brought down many bombers, the Allies could now send them in massive waves knowing they could replace the lost aircrews and planes. Still, if they eventually succeeded in starving Hitler's forces of gasoline, and forced diversion of labor and scarce material resources from weapons production into rebuilding and repair of plant and equipment, the preeminent effect of the bombing campaign was, by intent, to grind down the Luftwaffe as a fighting force. To this end, the Americans (and British) were willing to accept heavy losses of their own.[16] German interceptors, compelled to rise again and again to confront bombers numbering in the hundreds, were shot from the skies (or destroyed on the ground by marauding Allied escort fighters) faster than they could be replaced. German air power, and that of Japan in its turn, suffered most of all from losses of pilots. Neither country had planes or fuel to train capable replacements, and novices had little chance of surviving long enough to improve their skills in combat.

Yet the destruction of the Luftwaffe did not force Germany to capitulate:

> All the great expectations of air power as a distinct war-winning weapon that had been kept alive . . . by the single-mindedness of the bomber school were confounded by the fact that even the winning of the war in the air could not measurably reduce the time that it took to defeat the German armies in Europe.[17]

In the 1930s, aviators had claimed that strategic bombing would end the need for ground troops to fight and die in the mud as in World War I. It did not work out that way. Instead, strategic bombing became one more element in attritional warfare.

Once U.S. forces drew close enough for bombers to reach Japan from newly captured island bases, poor results from high-altitude strikes on industrial targets led Curtis LeMay, who had been sent from Europe to take charge, to order B-29s carrying incendiary warheads on low-altitude nighttime attacks against residential areas. "The destruction of 58 cities by fire-bombing . . . was to force the Japanese surrender . . . . completing a strategic programme that had eluded [the Army Air Forces] in Europe."[18]

The fire-bombing of Japan became the penultimate step in the redefinition of civilian populations as "strategic," a redefinition completed with the atomic bombs dropped on Hiroshima and Nagasaki.[19]

Precision Achieved

Since World War II, technical aids culminating in the satellite-based Global Positioning System (GPS) have made navigation to known target locations a simple task. Precision-guided bombs and missiles can now put a warhead in the pickle barrels that aviators claimed to be able to hit in the 1930s. Yet if these technical fixes overcame one set of limitations on strategic bombing, others remain, beginning with the need for good information on potential targets. Planners must know what is on (and under) the ground and where, something that remains subject to errors and uncertainties ranging from faulty intelligence to inaccurate or outdated maps.[20]

Those who believed that bombing, by itself, could be a potentially decisive weapon failed to gain a case in point even though Slobodan Milošević withdrew his forces from Kosovo in 1999 following 78 days of NATO bombing. The campaign took much longer than expected to show results and did little damage to Serbian military capability in Kosovo, where isolated targets such as tanks and artillery pieces were difficult for NATO aircrews to find and hit. It also made a long-standing concern of military professionals obvious to all: not only are some precision weapons too costly to permit much live practice, forcing military personnel to train with dummies and simulators, they sometimes cost far more than their targets. In the 1991 Gulf War, U.S. forces expended more than 17,000 laser-guided bombs, cruise missiles, and other precision-guided munitions, valued at nearly $2.3 billion. (Laser guided bombs cost about $30,000 each, cruise missiles several million dollars, "dumb bombs" $2000.) Not large compared to the total costs of the war, the sum is far from insignificant in the context of procurement spending. During the attacks on Milošević 's forces in 1999, cruise missiles were in short supply because production had been curtailed to conserve scarce dollars—and because the Air Force had shot off nearly half its inventory during three days of strikes on Iraq a few months earlier. As stocks of precision munitions declined, dumb bombs made up a growing fraction of warheads delivered by U.S. forces, eventually accounting for a substantial majority of the total (16,600 dumb bombs of 23,300 total).

While in other respects the 1991 Gulf War provided a near-perfect arena for display of precision weapons, clouds and fog, during the first ten days especially, forced many U.S. planes to divert to alternate targets or scrub missions entirely. NATO's 1999 air strikes in the Balkans provided a fuller picture of the capabilities of precision weapons given a more cunning adversary and more restrictive rules of engagement. Before the campaign began, NATO as much as promised not to send in troops, so that Serb commanders had no need to concentrate their forces for defense. Instead, they could disperse troops and hide their equipment. Aided by rough terrain, the Serbs showed a good deal more ingenuity in disguising potential targets than the Iraqis in 1991. NATO pilots found little to bomb other than scattered tanks, trucks, and artillery pieces, many of which turned out to be decoys. NATO could inflict little meaningful damage on Serbian military capacity.[21]

Average costs for precision weapons have been falling, in part because of inexpensive GPS guidance kits for converting conventional bombs. Nonetheless, the questions for mission planners remain much the same as in World War II. Compared to the overall costs of an air campaign, beginning with losses of aircrews and aircraft, how much harm can be done to the enemy's tanks and troops, command and control networks, economic infrastructure? If NATO aircrews cannot tell a Serbian tank from a dummy made of logs and camouflage netting, what do precision weapons bring to the task? If no one knows the locations of North Korea's nuclear weapons laboratories, how could they be destroyed? When the next Slobodan Milošević or Saddam Hussein installs antiaircraft radars on hospital roofs or missile launchers in schoolyards, should they be attacked? And if there are few targets of economic value in a country as poor as North Korea, how can bombing be considered strategic?

To qualify as strategic, a bombing campaign must seriously impair an adversary's military capacity and/or inflict sufficient damage on civilian or dual-use infrastructure (factories, bridges, powerplants) to bring the government to terms. Although NATO failed to substantially degrade Serbian military power in Kosovo in 1999, Milošević could be coerced. Attacks on economic targets in Serbia proper, much easier than finding tanks or howitzers moved as in a shell game from one hiding place to another, weakened Milošević's domestic support. As the bombing continued—and as NATO began to hint that an invasion of Kosovo might be forthcoming after all—Milošević's cronies grew restive, ordinary Serbs (e.g., those who found themselves out of work because of bomb damage) began to withdraw support, Russia's backing for its traditional ally weakened, and domestic rivals who could promise a return to something like normal life were emboldened. Even so, a more deeply entrenched ruler of a more ruthless police state would have had no reason to give in.

Like all weapons, precision bombs and missiles have their limitations. Many purely military targets can be hidden. Motor vehicles and aircraft can be stored underground, fuel and ammunition too. Since massed forces are vulnerable, adversaries will find other ways to fight. When they do, strikes on military targets are likely to devolve into the "tank plinking" and fruitless searches for mobile Scud launchers of Desert Storm, for Serbian tanks and artillery in Kosovo in 1999, for scattered Taliban forces in Afghanistan in 2001.[22] Disappointment with outcomes may then precipitate attacks on economic objectives. That brings a different sort of targeting problem. Mission planners must know enough about the adversary's industrial structure to identify vulnerabilities. The disintegration of the Soviet Union revealed how little understanding the Kremlin's own bureaucrats had of the economy they were supposedly managing; did U.S. intelligence do better? As the bombing campaign proceeds, finally, planners need accurate and timely damage reports as they put together each day's target lists.

Both rich economies (Nazi Germany) and poor (North Vietnam) have proved resilient under aerial assault. During World War II, the Allies managed to wear down, burn down, and cut off the war machines of both Germany and Japan from external sources of supply, especially of petroleum. But they did so chiefly through the more-or-less linear cumulation of pulverizing damage. The ideas of pre–World War II proponents of air power—that critical economic choke-points could be identified and destroyed, undermining the ability and/or will to make war—were simplistic. Under wartime pressures, labor can substitute for capital, alcohol for gasoline, scrapyard parts for new. Clever and capable technologists can find ways to substitute for, design around, or otherwise sidestep shortages of materials and components that lay observers might think irreplaceable.[23] Militaries have command and control systems that can be targeted. Not so for economies. They are not like arches that will collapse with extraction of the keystone; the closer analogy, for less-developed countries especially, is a heap of bricks that can only be flattened by pounding with big bombs in large numbers. Fission and then fusion weapons, with the destructive power to render precision delivery moot, long ago made it possible to contemplate such devastation. In the early 1950s, LeMay's Strategic Air Command could plausibly claim that a handful of B-52s, if they got through, would blot out the Soviet Union's cities. Yet indiscriminate attacks, even with conventional warheads, qualify as strategic only in the grossest sense.

A great deal has been written about the strategic bombing campaigns of World War II. Here the point is simply that the intersection of technology and doctrine, the locus of military innovation, has been subject to repeated miscalculation. Enthusiasts continue to embellish arguments for air power with metaphors such as "center of gravity," Clausewitz's term for the enemy's point of greatest vulnerability. NATO's bombing campaign in

Kosovo provides a recent illustration of how hard that center may be to locate, if indeed such a thing can be said to exist. Arguably suggestive in the 1920s and 1930s, the notion of a center of gravity, extrapolated from military forces themselves, has little meaning in light of current understanding of how economies function. And even in a purely military sense, the joint or "all arms" forces of recent years seem less likely to exhibit a unique center of gravity than the relatively compact ground armies and naval fleets of Clausewitz's day.

As an innovation, precision bombing was announced in the 1930s but not perfected until the 1970s. Laser-guided weapons made their public debut in the late stages of the Vietnam War. Yet if precision weapons then proved themselves technologically, that is irrelevant to the principal lesson, which is the long-running inadequacy of strategic bombing as *doctrine*. The precision weapons acclaimed since the early 1990s as revolutionary in themselves and also as a vital ingredient in the overarching Revolution in Military Affairs have a prehistory strongly colored by overstatement. At the same time, it is not at all obvious what precision weapons contribute to the peacekeeping missions and "operations other than war" likely to preoccupy the U.S. military in the future. For at least some of those missions, highly lethal weapons are useless.

Mistakes are part of innovation. But the double myth of strategic bombing—first, that intelligence could identify a short list of key targets that bombers would be able to locate and hit, and, second, that grave damage to the enemy's military power and will to fight would result—shows what can only be termed an extended failure to learn. In part because Allied planners had little sense of the actual workings of Nazi Germany's war economy, they miscalculated the damage their long-anticipated bombing campaign would inflict. The relative futility of precision bombing then served to rationalize, at least implicitly, the area bombing of cities. Deeply perverse in itself, that has had lasting consequences, not least as one precedent (among many) for terror attacks on civilian populations.

The Battleship: A Long, Slow Road to Obsolescence

The death of the battleship, and its afterlife, tell a more ambiguous story. During World War II aircraft carriers replaced battleships as the dominant means of projecting power at sea. The U.S. Navy made sensible choices in designing carriers during the interwar years. So did Japan, although not Great Britain. The U.S. Navy meanwhile learned little concerning the future place of the battleship. Exercises could not be definitive. And the lessons of engagements in the Pacific following Pearl Harbor were, like so much else in military innovation, less than clear-cut. Once the war was over, however, and the message reasonably plain, many naval officers were reluctant to accept that the battleship no longer had a claim on any vital mission.

Even as the Army Air Corps in the 1930s prematurely embraced the unproven technology and doctrine of strategic bombing, naval aviators sought, in sharply contrasting fashion, to understand through war games and fleet exercises what aircraft carriers could contribute to power projection at sea. British authorities declined to store aircraft on deck where they would be exposed to enemy attack (and North Sea weather); instead they built carriers with armored flight decks beneath which to shelter planes.[24] The U.S. Navy made an opposite choice, not only accepting the risks of parking planes on the flight deck but electing to construct those decks of wood for rapid repair of battle damage. The British, limited by available space on their hangar decks, had to make do with fewer planes—around 60, compared with 90 or more for the large fleet carriers of the United States.[25] In accepting that planes would have to be shuttled between hangar and flight decks, the British also had to accept slower rates of launch. These decisions seriously impaired the operational effectiveness of the Royal Navy's carriers. At the same time, their armored decks provided badly needed protection against kamikaze attacks during the late stages of the Pacific war, when those U.S. carriers yet to be fitted with deck armor proved dangerously vulnerable.

During World War I, the battleship-heavy British and German fleets had mostly stayed at home, stalemating one another. By the 1930s, it was clear that airplanes threatened surface ships at least as much as the mines and torpedoes commanders had earlier learned to fear. As yet, however, there was no conclusive evidence that battleships were on their way to obsolescence. No other vessels carried nearly as much armor. Their hulls subdivided into watertight compartments, decks thickly plated to deflect or arrest downward plunging shells, fitted with extensive fire-fighting and damage control systems, battleships were the most robust of warships by far. Lesser vessels might be vulnerable to dive bombers, but only land-based planes like the B-17 could lift a gravity bomb big enough to damage a battleship and they had little hope of hitting a maneuvering vessel from altitudes great enough for their bombs to reach the high speeds necessary to penetrate a battleship's armored deck. Only torpedo planes could threaten a battleship from the air.

Battleships had to be big simply to float the monstrous guns housed in their central citadels, thickly armored so that the big guns could continue firing even if bow or stern were blown away. Fleet carriers were just as big, but could be neither heavily armed nor armored without impairing their mission: they needed interior space and flotation to accommodate as many planes as possible and the spare parts, aviation gas, and munitions to keep them flying. Like other warships, carriers bristled with antiaircraft guns; but as prime targets for the enemy they would be hard pressed to fend off swarms of attacking planes or warships in the absence of air cover provided by their own fighters. When darkness, weather, or sea conditions shut down flight operations, carriers were vulnerable and could only depend on escort vessels for protection or else try to hide.

For a time, then, battleships seemed a still-useful part of the fleet, bulwarks to help shield carriers from the guns and torpedoes of the enemy's surface ships and contributors to the withering curtains of antiaircraft fire that supplemented air cover in protecting the entire battle group from the enemy's attack planes. From 1937 to 1940, the U.S. Congress authorized construction of 17 new battleships. Yet the battleship was a static technological system at a time of rapid ongoing advances in aviation and airborne radar, which together were transforming blue-water warfare. Land-based planes, faster and less vulnerable than the Navy's long-range flying boats, took over much of the work of search and surveillance at sea (carriers embarked relatively few reconnaissance planes, which came at the expense of fighting power). Long-range land-based planes fitted with sensitive radar equipment made it steadily more difficult for ships to vanish in the nearly empty ocean, as they had been accustomed to doing for centuries. Fog and rain still promised refuge, but a commander could no longer expect to slip off and disappear if outgunned or outnumbered. Battleships were no exception. High-value targets with thousands of men aboard, they remained a priority second only to aircraft carriers for an enemy able to maneuver a submarine, destroyer, or plane within torpedo range or amass sufficient firepower to risk a surface attack.

Battleships were feared because a single projectile, weighing more than a ton, from one of their big guns could blow a lesser vessel literally in two. Their primary missions had always been to fight enemy battleships and sink any of the opponent's other ships they could locate and overtake. Other weapons platforms now performed those tasks better. No heavy guns fired during the Battle of the Coral Sea in May 1942, when the Allies checked Japan's southward advance toward Australia; air strikes accounted for all losses on both sides. Following the Battle of Midway a month later, the United States dropped its plans for a new class of battleships in favor of more carriers.

The lesson came quickly and perhaps fortuitously. The Navy's carriers had been at sea when Japan struck Pearl Harbor. The core of its Pacific fleet otherwise decimated, the United States had no choice but to rely, at least until more battleships could be built, on air power. Within two years it was plain that aircraft carriers and submarines had become the dominant weapons of naval warfare.[26] With the conclusion of the island-hopping campaign in the Pacific, the Navy's battleships could not even claim shore bombardment as an essential mission. Their big guns had never exhibited much accuracy at long ranges and naval leaders who closed their eyes in later years to the inability of battleships to find their targets matched the earlier self-deception of advocates of precision bombing:

> [I]n April 1968, . . . the modernized USS New Jersey was recommissioned and sent to Vietnam. During her single cruise off the coast, New Jersey fired 5,688 16-inch shells and succeeded in inflicting 113 confirmed enemy deaths and blowing up a small island in the process.
>
> In 1983–84, off the coast of Lebanon, the New Jersey sought to suppress hostile Syrian and Druse militia artillery positions in the hills. Lacking forward spotters, the big guns proved wildly inaccurate, on one occasion decimating a herd of goats rather than the 21 mm antiaircraft guns against which they were targeted.[27]

During the interwar years, the Navy compelled aviators to earn a place in the fleet. Doctrine and design choices for aircraft carriers were based on empirical evidence. That evidence was far from complete, but it provided sufficient understanding to enable the U.S. Navy to beat back the Japanese fleet. The Army, for its part, conducted no meaningful evaluations of strategic bombing; after isolating the Air Corps, the high command paid little attention (except to deny it funds—President Roosevelt intervened personally in 1938 to enable the Air Corps to purchase big bombers). While the Navy's receptivity to air power owed something to the service's desire to forestall Army aviators who would have been happy to take over all military aviation, the openmindedness of the admirals still contrasts sharply with myopia in the Army. It also contrasts with shortsightedness on other issues. Most egregiously, naval leaders refused, well into the submarine campaign in the Pacific, to credit reports of defective torpedo fuzes, failing even to order tests that would verify or refute those reports.[28] And when it came time to consider the postwar future of the battleship, the Navy did not probe deeply, even though many of its own officers deemed the big ships good for nothing but targets for gunnery practice.[29] The common element: failure by high-level authorities, military and civilian, to insist on realistic, evidence-based evaluations.

Prior to the establishment of the Office of the Secretary of Defense in 1947, such assessments could only have come from the military itself.

Civilian officials had hardly any means of arriving at independent findings. Should Congress or the White House wish to question the claims of the generals and admirals, or the service secretaries who spoke for them, they had little recourse but to convene some sort of special advisory board or panel. While these were not uncommon, if unable to steer the study to conclusions they favored, the services, then as now, could expect to stall until a new administration took office.

Civilian Nuclear Power: Investment Outpaces Learning

In the civilian economy, market competition snuffs out innovations when capability relative to cost fails to improve over time. Magnetic bubble computer memory never displaced conventional integrated circuit memory because its cost/performance increases could not keep pace with those of dynamic random-access memory chips. In the 1980s, many U.S. manufacturing firms purchased industrial robots for their factories expecting reduced production costs and improved quality. When gains proved smaller than expected, robots did not disappear; but they did not diffuse at the rates that had been expected earlier.

When innovations such as industrial robots are new, prospective users may feel they have no choice but to invest, even though there is little basis for predicting returns. Better to risk disappointment than take the chance that competitors will get a head start down the learning curve and perhaps be able to remain in front for years. Civilian nuclear power shows what can happen when firms plunge into a new family of technologies requiring heavy investments in the absence of an experience base and without allowing time for experimentation and learning.

The military origins of commercial power reactors make up only a minor strand in this story. In the 1950s, the U.S. government sought to display America's technological prowess by showcasing the "peaceful atom." Washington hoped to gain political ground among nonaligned nations and a bit of moral stature in the aftermath Hiroshima and Nagasaki, while also promoting domestic and foreign sales of reactors, generating equipment, and construction management services. At first, the Atomic Energy Commission (AEC) had no success in persuading domestic utilities to invest in nuclear power. Given vast U.S. coal reserves, all projections showed that fossil-fuel plants would be able to supply electricity at lower costs for decades to come. Even Lewis Strauss, who as AEC chairman foresaw nuclear electricity becoming "too cheap to meter," was careful to put the time a generation ahead.[30] Subsidies would be needed. In 1954, Duquesne Light Company responded to an AEC solicitation by proposing a demonstration plant to be built at Shippingport, Pennsylvania. The federal government agreed to pay the bulk of the $55 million cost, retaining ownership of the reactor and selling steam to Duquesne to drive the company's turbo-generators.

The AEC was already deeply involved in light water reactors for the Navy (chapter 5). Indeed, this was the only reactor design with which the AEC, the Navy, and their primary contractors, General Electric (GE) and Westinghouse, had meaningful experience; other reactor configurations existed only as concepts.[31] With a viable proposal finally in hand, the AEC, prodded by Congress's Joint Committee on Atomic Energy, saw no reason to look beyond light water technology. The Commission, not without qualms in view of his reputation and management style, asked Rickover to take charge of the project. He based the Shippingport reactor on a design originally intended for an aborted nuclear-powered aircraft carrier.[32]

The Shippingport plant, the Navy's only direct engagement with commercial nuclear power, went on line at the end of 1957, a technical success. With Washington proffering a host of direct and indirect subsidies, enthusiasm for nuclear power grew among previously skeptical utilities. Military R&D and procurement had created a substantial technology base, much of it within GE and Westinghouse, both of which evidently set prices on early units well below cost in efforts to establish position in what they believed would be a lucrative long-term market. Congress provided insurance against liability through the Price-Anderson Act, the federal government paid a portion of the capital costs for a number of post-Shippingport plants, and also took on the costly, contentious, and still unresolved problem of dealing with spent fuel and radioactive wastes.[33]

The investment wave crested during 1966–1967, when domestic utilities ordered more than 60 nuclear powerplants. By 1967, the generating capacity of the plants on order exceeded the capacity of those that had been completed by 25–30 times.[34] Even the smallest of the new reactors were much bigger than any built for the Navy, and those on the drawing boards were bigger still, although utilities had yet to gain meaningful operating experience with nuclear power, or indeed much construction experience. Normally conservative utility managements committed large sums without waiting for actual cost figures to come in.[35] Schedules slipped, sometimes by years, costs rose far beyond the investment levels approved by regulatory bodies, anticipated scale economies proved elusive. It took another half-dozen years to absorb the lessons. Then disillusionment set in. Although 100-plus nuclear powerplants continue to generate about one-fifth of the nation's electricity, the last order for new construction not subsequently canceled was placed five years before the 1979 Three Mile Island accident.

Federal policies, stemming ultimately from East-West rivalry, bear much of the blame for the nuclear power muddle. The AEC's monopoly over bomb-building extended to nuclear electricity. The Commission provided vocal and financial support but its technical support was deficient. Preoccupied with building up the nation's stockpile of nuclear warheads, the AEC neglected its nondefense R&D programs, which, starved for funds and marked by uncertainty of purpose and organizational disarray, did little to address the practical issues of reactor design that later emerged as paramount for utilities, such as passive safety.[36] Rickover and the Navy had chosen light water reactors because they could be made compact enough to fit within the hull of a submarine. Light water was not obviously best for commercial power, but the AEC did not explore alternatives. The government's financial subsidies distorted investment decisions by utilities that lacked the technical expertise to conduct their own evaluations and were too ready to accept the claims of the AEC, equipment suppliers, and the engineering and consulting firms that specialized in powerplant design and construction. Because coal prices remained low and air pollution regulations would not begin to bite until the 1970s, utilities had no compelling reason to purchase nuclear powerplants. Yet they did. The reasons for the investment bubble continue to resist full understanding. Competitive pressures sometimes induce firms to invest in new technologies without any basis for predicting returns—e.g., for business investments in robotics or some information technology projects.[37] Yet no such pressures existed on the regulated utilities of the 1950s and 1960s. The absence of competition, on the other hand, meant that there was little to discipline decisions by utility managers, who were accustomed to regulatory approvals of rates set to cover whatever sums they chose to invest.

The nuclear power bubble shows what can happen when public policy short-circuits experiential learning. If utilities had proceeded with less haste, waiting for lessons to come in from building and operating early nuclear plants, diffusion would have been slower, the mid-1960s investments would have resembled a ripple rather than a wave, and the subsequent backlash against nuclear power would perhaps have been muted. That backlash was sharp enough to stall R&D on alternative reactor designs that might have led to technologies more acceptable to the public and would perhaps be available today for generation of electricity without release of greenhouse gases, a problem that had little visibility until the 1980s.[38]

Feedback from users into military innovation must be deliberately managed, else there is likely to be no feedback at all. The task falls mainly to the Pentagon; defense firms participate when they believe it in their self-interest. Lacking market-mediated feedback, the process of innovation is prone to breakdowns resembling the nuclear power case. The Air Force had no more need of the B-1B bomber in the 1980s than utilities had of nuclear electricity in the 1960s. We might say, with little risk of oversimplification, that politics overrode policy in both cases. For the B-1B, the political dynamics centered on inter- and intraservice rivalries and the Reagan administration's fiscal and budgetary machinations. For nuclear power, Cold War geopolitics underlay decisions taken by the AEC and Congress. In both cases, insiders controlled the policy debate. They relied on secrecy, appeals to specialized professional knowledge, and obfuscation to get their way.

CHAPTER 10

GENERATION OF VARIETY AND SELECTION OF INNOVATIONS

[I]n dealing with capitalism we are dealing with an evolutionary process.

The fundamental impulse that sets and keeps the capitalist engine in motion comes from the new consumers' goods, the new methods of production or transportation, the new markets, the new forms of industrial organization that capitalist enterprise creates.
—Joseph A. Schumpeter[1]

This chapter continues the comparison of military with commercial innovation. In the framework adopted, economic forces act on commercial innovations through "environmental pressures," by analogy with evolution in nature. The analogy is partial because, while genetic mutations are random, their technological equivalents stem from deliberate human action. Innovations likewise diffuse and find new applications as a result of conscious decisions by inventors, entrepreneurs, business firms and their customers, and governments. Natural selection is driven by the blind workings of genetics, technological innovation by discovery, information transfer, and learning. Despite these differences, the analogy is powerful. It keeps the focus on competition and on the characteristics of the settings, market and nonmarket, in which competition takes place.

Any evolutionary picture of innovation must include two sets of mechanisms, those for creating technological variety and those for winnowing variety through selection. While variety is sometimes associated with R&D and invention, it has other possible sources, such as marketing and design. Mechanisms for reduction in variety through selection differ more deeply between military and commercial innovation than do mechanisms for generation of variety. Selection of commercial innovations takes place through the play of market forces, which leads to the survival of some products and processes and the stillbirth of others. Firms may try to override these forces but seldom succeed, although they can sometimes shape and control them. For military technological innovations, reduction in variety is primarily a matter of political and bureaucratic choices. These resemble the choices made *within* a firm, as contending factions struggle for resources to pursue favored ideas, product strategies, and indeed lines of business. Once the firm has made its choices, market competition sorts through them, in the process sorting more successful managers and firms from less successful. For military technological innovations, selection pressures of comparable intensity are activated only in wartime.

While commercial innovation is fast-paced and unrelenting, military technological innovation is episodic. No matter how many ideas may be generated, only a few can be implemented in practice, because new weapons programs are so costly. Since the 1960s, moreover, differences between military and commercial innovation have deepened, for two primary reasons. Military and commercial technologies have diverged, at both system and component levels. Second, the manufacturing industries in which formal R&D is a prime source of innovation have contracted relative to service-producing industries in which innovation is less obviously technological in character.

Technological Evolution

Sources of variety feeding into commercial innovation include research and development, design and development, marketing, and business planning and strategy (table 8).[2] Some of the parallels in military technological

Table 8   Evolutionary Mechanisms in Innovation

	Commercial Innovation	Military Technological Innovation
Generation of Variety	R&D/D&D, market-driven entrepreneurship, user-induced demand.	R&D/D&D, political and bureaucratic entrepreneurship, user demand.
Selection: Winnowing and Reduction in Variety	Market outcomes.	Political and bureaucratic decisions; warfighting outcomes.

innovation are close (R&D and D&D), others remote (in terms of objectives, national security strategy has little in common with business strategy).

Depending on the industries in which they compete, firms must innovate more or less constantly if they expect to survive, prosper, and grow. In some industries, "innovation" amounts to little more than superficial product differentiation. Retail banks offer roughly comparable packages of services (with possible exceptions for Internet banking, where considerable uncertainty still attaches to the evolutionary trajectory), while microelectronics firms literally race to be the first to market with technologically advanced products such as integrated circuit (IC) chips with features that respond to the preferences of major customers. Those preferences change over time. Recently, for example, purchasers of ICs for portable electronic products have wanted chips with low power consumption to reduce heat generation and to offer their own customers extended battery life. In industries including wireless telephony and automobiles, product strategies combine technical advances with design attributes intended to appeal to fashion-conscious consumers. Firms seek to differentiate their products without moving too far from mainstream preferences. In all these cases, marketing considerations combine with technological developments to shape the final product.

Innovation and technical change themselves alter market characteristics and selection pressures. The market for ICs behaves differently today than in the 1960s, when production for military and space systems accounted for the great majority of sales. Deregulation has transformed the market for air travel. Before 1978, with fares fixed on most routes, airlines competed primarily on quality of service, which became a spur to innovation. Barred from fare-based competition, airlines sought to attract customers by flying the latest planes available. They backed up their orders with advance payments which provided cash that aircraft manufacturers could direct to design and early production. Today, with competition based almost entirely on price, airlines seek to reduce their operating costs so they can maintain profitably with low fares. Then and now, airlines have valued the fuel efficiency of the engines fitted to their planes, since lower costs boost profits; but economical operation has become far more important since deregulation.

Part of the task of innovating firms is to understand the behavior of the markets in which they compete and properly interpret the signals generated in those markets. In the late 1960s, a small company, Docutel, commercialized the automatic teller machine (ATM) and began selling computer-based ATMs to retail banks as a way of cutting labor costs. Banks that bought the new machines (not many, at first) planned to replace tellers in their branch offices and close some of those offices. Soon, banks began to see ATMs as a potential source of "strategic" advantage. Because customers could make withdrawals and deposits around the clock, banks had stumbled upon a fundamentally new service product. A widespread ATM network enabled a bank to offer, and to advertise, greater convenience than its rivals, differentiating itself in an industry that otherwise offered commodity-like products. Competition had switched from cost-cutting to market promotion. As ATMs spread, other suppliers joined Docutel in designing and selling them. Banks, meanwhile, recognized that they could extend their reach by joining together to create shared ATM networks. This in turn made ATM services a commodity, like other retail banking products, preventing any bank from establishing a sustainable advantage. Banks eventually reversed their old strategy of closing branch offices to cut costs and began to open new offices in selected locations to enhance their neighborhood presence and offer services, such as home mortgages, that ATMs could not provide. Over time, one set of selection pressures acted on banks, another on ATM manufacturers. Banks that were alert and responsive to market dynamics and customer behavior prospered; those that lagged found themselves swallowed in mergers. As the market for ATMs grew, suppliers focused on dependability and low maintenance. Once customers had come to trust ATMs, opportunities opened for related innovations in unattended delivery of services such as credit card–operated gasoline pumps and, more recently, electronic airline ticketing.[3]

Applications of the Global Positioning System (GPS) offer a distant yet revealing parallel. During the 1991 Gulf War, GPS provided substantial advantages to U.S. troops, who could quickly, easily, and accurately determine their own location and keep track of the locations of friendly forces even in featureless desert terrain. Since then, widespread sales of low-cost GPS receivers have vitiated this source of advantage. Although the U.S. Department of Defense (DOD) can degrade unencrypted GPS accuracy (and jam signals selectively), a growing list of civilian applications, some of them safety-critical (e.g., aircraft navigation), mean that all sides in future conflicts are likely to have access to GPS signals good enough for many military purposes. In effect, GPS, once a proprietary service provided by DOD to the U.S. military and its allies, has, like ATM services, become a commodity, available equally to hunters, hikers, and terrorists.

In the short term especially, market signals can be puzzling. Firms may realize too late that they have missed a twist or turn by months or years. Even so, the task is easier than evaluation of prospective military innovations. Because businesses compete every day, they get readings every day in the form of sales figures, warranty claims, the stock market's valuation of their tangible and intangible assets. They can track performance by line of business, search out improvements through activity analysis and benchmarking, survey their customers and talk to suppliers. Military organizations employ similar tools when they can, and indeed have contributed methods such as operations research, the family of mathematical and statistical techniques that after World War II became a building block for "management science." But meaningful readings of performance come only from warfighting, and after the war is over analysts, historians, and participants with interests at stake put forward interpretations that, as in the aftermath of the Vietnam War, may be debated for generations.

Emulation is a primary mechanism for diffusion of innovations. Banks and military organizations alike observe and react to the choices made by their peers, seek to understand, copy, and improve upon successes and to skirt pitfalls revealed in hindsight. In the civilian economy, innovations spread, combine, mature, and multiply. As in nature, technological evolution is full of blind alleys, detours, dead ends, and more than occasional freaks. Over time, selection winnows down variety. Some branches of the evolutionary tree grow rapidly; others wither and die, finally to be snapped off by the Schumpeterian gale of creative destruction. Centrifugal compressors disappear from jet engines; axial compressors remain. Analog computers, widely used a half-century ago to solve some of the most formidable technical problems of the time, were eclipsed in the 1960s by digital computers, which had the flexibility to attack a wider range of problems. As engineers and scientists turned to digital machines for ever more complicated calculations, their efforts spurred advances in modeling techniques, algorithms, and programming languages; these combined with rapid gains in hardware performance to make digital computing steadily more attractive. Already, analog computers have been nearly forgotten.

Variety

Technical attributes, hence technological variety, can be represented in several ways. For manufactured goods, attributes are embodied in products themselves and in the blueprints, databases, and process sheets that specify the data, information, and knowledge necessary to replicate the design. Standardized services, as in retail banking or fast foods, are commonly produced according to company-specific routines (e.g., scripted customer interactions) that employees follow, sometimes with the aid of computer systems. For nonstandard services tailored to the individual customer, such as those provided by investment banks, product attributes are determined in the course of delivery; often the customer participates, coproducing the service. Whereas in manufacturing industries, design and production are separate and distinct, for all except the most standardized services they are combined and interdependent.[4]

Empirical studies and theory have not yet yielded a widely accepted framework for understanding and analyzing innovation. The many historical accounts of invention leave the underlying processes largely a mystery. Case studies of innovation itself, in the sense of commercialization, suggest that almost anything entrepreneurs and firms do sometimes contributes. Most of the theoretical treatments so far available are limited to innovations that originate in R&D, for reasons of analytical tractability (R&D spending and employment of R&D personnel are the only widely available measures of inputs to innovation), even though empirical studies in the 1960s showed that innovations have many sources other than R&D.[5] Shopfloor learning by factory workers, for example, has contributed to the package of innovations that came to be known as lean production in the auto industry. We saw earlier that the microprocessor resulted from D&D rather than R&D. Many firms credited with innovations in service industries, finally, report little or no spending on R&D. They simply do not think of innovation in those terms.

If research in the sense of purposeful search for new knowledge is not a necessary prelude to innovation, it is nonetheless well understood as a result of extensive study by historians and sociologists of science and is thus a logical starting point for exploring both generation of variety and selection. For that reason, the discussion below centers on R&D. This by no means implies endorsement of linear or pipeline models of innovation, which have been thoroughly discredited.[6]

Research aims at discovery of knowledge that can help resolve puzzles ranked high in significance by specialized disciplinary communities. These communities, their members known by their accomplishments, pass judgment on proposed additions to the knowledge base, bestow rewards for discovery, and establish priorities for further research. Academics predominate in the sciences, less so in engineering. Through the exchanges of insight and information that are part of the everyday work of science, the community reaches implicit conclusions, at first tentative and then with increasing certainty. In doing so, it confers rewards more or less proportional to the community's assessment of value. Scientists and the small minority of engineers who engage in research vote with their citations and through the directions they chose for their own work: findings that stimulate efforts at extension, helping set the overall direction of a field, earn high accolades. Suspect results, those that fail to pass preliminary screens (trivially, findings that cannot be reproduced), do not enter the knowledge base. Consensus emerges; no authority imposes it, nor could any do so.[7]

Only a handful of people may be competent to fully grasp the significance of prospective additions to the stock of scientific and technical knowledge. Many disciplinary communities are small and inbred, as well as self-chosen. Even so, their judgments are self-correcting because the process is open to those with sharply conflicting perspectives. Scientists, no matter how opinionated, contentious, or famous have no power to force their views on others.

Within these specialized communities, research goals are widely understood: explicating the mechanisms of high-temperature superconductivity in terms of the quantum theory of solids and observed microstructural features; synthesizing materials that will conduct electricity with zero resistance at higher temperatures and in the presence of stronger magnetic fields; fabricating conductors in long lengths for practical applications. While research objectives remain unstated in many fields, in others such as genome mapping and high energy physics, where advances often depend on collaborative projects involving dozens or hundreds of research workers and costly experimental equipment, study groups or committees negotiate explicit priorities for presentation to sponsoring government agencies and foundations.

Research directions in defense are chosen in much the same way as for any other work. After World War II, competition among both agencies and research groups within the relatively new and rapidly growing military system of innovation enhanced variety. New ideas could get a hearing, if not by the Navy then perhaps the Air Force or Army, if not by one of the services perhaps by the Defense Advanced Research Projects Agency (DARPA), which supported work the armed forces sometimes prefered to avoid.

Any funding agency must respond, to greater or less extent, to political forces. Those forces may be stronger in the Air Force Office of Scientific Research (AFOSR) than the National Science Foundation (NSF), but they are hardly absent in the latter. They are strong, too, in the National Institutes of Health (NIH), pressured by Congress and "disease lobbies" to support particular fields of research. At the same time, the principles established by Vannevar Bush's Office of Scientific Research and Development (OSRD) during World War II continue to underlie operating practices in many of the federal government agencies that finance research (agriculture, which has its own traditions predating OSRD, is the major exception). Bush believed that "Scientific administration was best managed . . . with responsibility weighted toward the bottom of the organization and in the hands of the men in closest touch with the projects themselves." As a result, wartime programs were "cobbled from tens of thousands of separate bargains stuck by individual scientists, entrepreneurs, and officers."[8] After the war, the somewhat awkward mix of top-down political guidance and bottom-up implementation that Bush had implanted in OSRD migrated along with many of OSRD's administrative and contracting procedures to newly established R&D agencies, military and civilian alike, in other parts of government (chapter 5).

Within these agencies, funding decisions continue to reflect the interplay of bottom-up judgments of scientific or technical promise by working-level employees, some of whom are themselves deeply engaged in research, and top-down priorities set at higher levels in government, often by political appointees. NSF relies heavily on program officers who rotate through for tours of a year or two from academic positions. Many of NIH's scientific administrators maintain their own intramural research programs. AFOSR administers external research contracts solicited and selected by working scientists and engineers at the Air Force Research Laboratory. Through multiple channels, political and bureaucratic forces influence research directions; rarely do they override scientific opinion (restrictions on federal funding for stem cell research are the exception), but they shape it constantly. AFOSR and DARPA, like NSF and NIH, listen to the congressional committees that appropriate their funds. They also listen to outside scientists, whose views they solicit individually and through workshops and conferences. Politics exerts far less influence over DOD's research than over design and development. Most research is too far removed from weaponry to attract the attention of high-ranking officers who wield decision authority.

Uncertainty

In the world of research, the marketplace of ideas eclipses the marketplace of commerce. Scientists compete for funding, for the best graduate students and postdoctoral fellows, to publish in prestigious journals and attract citations that validate their work. The rewards do not depend on possible applications, if any, since these are likely to be far off and indistinct. While prospective profits do motivate some research in fields such as high-temperature superconductivity, and potential applications have considerable significance in biomedical science and the engineering sciences, the incentives tend to be for relatively generic results that do not presuppose particular applications. The chief reason is uncertainty. By definition, the outcomes of research cannot be predicted. If they could, the research would not be considered "basic": extension of disciplinary knowledge in predictable directions is more commonly associated with applied research. Scientists and engineers often have strong convictions concerning the likely long-term outcomes of their work, and its value (and perhaps equally strong opinions concerning the research of their rivals), but these views can only be matters of belief and intuition, "intellectual mechanisms which we do not know how to analyze or even name with precision."[9]

Uncertainty is the reason why something less than 5 percent of business-funded R&D supports work classified as basic.[10] Lacking reliable grounds for prediction, returns on investment cannot be usefully estimated. Businesses know there is a high probability that if they fund such research the results will not benefit them, while perhaps benefiting others, possibly competitors or firms and industries that are not even in existence yet. As a distinct industry, microelectronics began to take shape a decade or so after the invention of the transistor at AT&T Bell Laboratories. Many leading manufacturers of IC chips were founded well after the commercialization of the integrated circuit by Fairchild Semiconductor and Texas Instruments (TI). And while design and development projects at Xerox's Palo Alto Research Center underlie many familiar features of today's PCs (the mouse, graphical user interfaces), Xerox never succeeded in capturing large rewards from its innovations in computing and remains primarily a manufacturer of imaging equipment.

After World War II, policymakers came to understand that society as a whole could expect large returns from a broad and deep research portfolio, while private investors, lacking confidence that they would capture a share of the eventual returns commensurate with their expenditures, had little reason to pay the bills. Policymakers acted on this understanding well before economists provided, during the 1950s, the analytical justification in terms of uncertainty sketched above.

If and when prospective applications begin to appear, firms have stronger incentives to invest. Most corporate spending supports work that is closer to D&D than R&D. (Firms in the automobile industry spend more on R&D than firms in any other industry; almost all of the money goes for design and engineering of new models.) Both R&D and D&D celebrate creativity, cleverness, elegance, and other evocative if ethereal terms associated with new ideas and discovery. At the same time, the context for design and development is the search for ideas that will yield profits through new products, new processes, or indeed entirely new lines of business. In this, D&D differs fundamentally from research. Most D&D, perhaps needless to say, is not creative in any deep sense. Almost all new product designs are derivative, variations on a theme (this year's PCs or minivans, as opposed to the first of their kind). In science, this kind of extension is viewed as work for dullards (relatively speaking). Engineers, on the other hand, are trained to think in terms of what is, or will be perceived as, "new and improved" in the marketplace and managers are rewarded for guiding D&D to products that will cover their costs and go on to generate the profits necessary to support further innovation.

Businesses generally require technical and product-line managers to justify budget requests on the basis of projected returns. No one expects these projections to be accurate, only transparent, based on uniform procedures and explicit assumptions intended to stop competing managers from rigging their proposals too blatantly. Relative rankings (e.g., on a net present value basis) combined with managerial judgment then guide allocations of funds. Businesses accept uncertainty. Recognizing that some of their investments will not pay off, they maintain portfolios that they hope will generate enough winners to support the entire enterprise.

Selection

There are two sets of selection mechanisms to consider: those that assign value to research results and those that select among innovations themselves, ideas reduced to practice rather than disembodied knowledge. Selection among candidate additions to the knowledge base takes place much as does choice of directions for research, through informal evaluation by communities of experts who develop or make use of related knowledge. Acceptance of prospective military innovations, by contrast, depends on persuading decision-makers, some of whom (e.g., in Congress) may have little grasp of either technology or military affairs.

DOD-sponsored research, with exceptions for projects classified secret and some work conducted in the military's own laboratories, must pass through much the same set of screens as any other work.[11] Lysenkoism could survive for a time in Stalin's Russia and quacks selling cures for cancer may deceive some people; but within the community of science errors are selected out promptly, as illustrated by the discrediting of cold fusion in the 1980s.

For technical methods, selection differs somewhat. Expertise is still needed—engineering science can be highly esoteric—but the relevant communities tend to be larger, with greater representation by industrial employees. The primary criterion becomes utility rather than truth. Technical methods that provide help with practical problems—techniques for predicting radar returns from stealthy aircraft, mathematical models of traffic congestion on a busy freeway—will be accepted and adopted even if they violate aspects of physical reality. That is not the case in science. For example, mathematical models of cracks play a central role in fracture mechanics, enabling predictions of failure in airframes, ships, and other structures.

In many of these models and methods, the calculated stress at the crack tip becomes infinite. That is not only physically impossible, it is quite unrealistic: when a crack extends, material separates in precisely this region, leaving a free surface at which the stress must be zero by definition. Accuracy close to the crack tip would seem to be essential for prediction; yet as it happens, these methods give good results (for reasons not totally understood) and have become widely accepted. In technical practice, expediency matters more than truth to nature.

For innovations themselves, winnowing and reduction in variety take place more-or-less spontaneously, with success or failure registered in the marketplace. Businesses try to manipulate customers, modify selection pressures, and shape market outcomes through advertising, pricing strategies, and monopolization. Sometimes they succeed. Most observers considered Sony's Betamax videocassette format technically superior to Matshushita's competing VHS standard, which won out. Government subsidies boosted investment in nuclear power, with early choices by major purchasers contributing to lock-in effects: when big utilities purchased nuclear power-plants, smaller utilities followed along, some of them more-or-less blindly. On the other hand, DOD had little success in establishing its Ada computer language even in purely military applications.[12] Finally, pioneering firms sometimes fail while their innovations survive. Docutel disappeared a few years after commercializing the ATM, the first manufacturers of PCs, such as Altair, did not last long, and De Havilland, which produced the first jet-propelled passenger plane, the Comet, could not survive following the crashes of three of these planes in 1953–1954 as a result of fatigue cracks that engineers now guard against with the tools of fracture mechanics.

The technological variations to which markets assign high value may not at first be evident. Once Intel introduced the microprocessor, the question became who, other than the original customer, Busicom, would buy the new chips. At first, Intel's marketing department was pessimistic, projecting sales of only a few thousand units per year, a faux pas much like IBM's two decades earlier in putting worldwide demand for large computers at two dozen or so.[13] In the early 1970s, few people foresaw that microprocessors would find widespread applications in industrial process control, much less PCs, which had not yet been "invented." Nor did market studies in the late 1970s indicate that the first major market for PCs would be as replacements for typewriters in offices, perhaps because this application depended on innovations from a different set of firms, those that supplied software. Office automation via the PC soon got a further boost from the introduction of spreadsheet programs, which enabled businesses to replace the adding machines and rotary calculators used by armies of clerical workers, bookkeepers, and accountants with the same PCs used for typing. Almost overnight, companies such as Friden, which in the early 1960s had begun developing hard-wired electronic calculators to supplement its line of electro-mechanical office machinery, saw the bottom drop out of their core market. In all these cases, the workings of selection processes became evident mostly in hindsight; like Friden, many firms misjudged the opportunities that opened as a result of cascading innovations in the underlying technologies of digital electronics.

Engineering design is in considerable part a search for evaluative criteria. What do users want or need (as opposed to what they think they want or say they need)? How do they balance design attributes that trade off against one another? Intel managers knew that future microprocessor sales would depend on understanding customer applications and what those implied for purchasing preferences. When the company began work on its 386 family, several generations after the pioneering 4004, the lead technical and marketing specialists spent six months simply visiting customers to explore alternative design choices such as instruction sets (the list of commands the processing unit can execute).[14] Digital Equipment Corporation, for a time the world's second largest computer manufacturer after IBM, devised a strategy for the PC market around a very fast proprietary chip, its Alpha microprocessor. Digital believed a speed advantage would win substantial sales. The Alpha chip won a place in the Guinness book of records as the world's fastest, but could not reverse the firm's sagging fortunes.[15]

Once a fundamental innovation has taken root, evolutionary pathways begin to branch. Often, as in microelectronics, they do so repeatedly. Sensing future profits, TI and Fairchild Semiconductor each developed integrated circuits in 1958–1959. The companies did so independently, taking quite different technical approaches. Research, in the form of a relatively systematic search for new knowledge, contributed much more to Fairchild's invention than to TI's, which fits the mold of inspired invention more closely.[16] Their undertakings were responses to the same stimulus. Both DOD and the National Aeronautics and Space Administration (NASA) wanted small, lightweight, rugged, and reliable electronic components to replace vacuum tubes and discrete transistors. For some years, the idea of combining several transistors on a single chip had been "in the air," but no one had been able to build such a device. In the mid-1950s, the Air Force initiated an R&D program on three-dimensional "molecular electronics." The concept proved overambitious and the Air Force got little direct benefit. But the push for some sort of densely packed electronics module signaled the strength of latent demand for miniaturized solid-state components. Neither TI nor Fairchild received R&D funding from the Air Force, but both firms recognized that sales and profits would follow if they could produce devices that met the needs of DOD and NASA. (This was exactly the motivation that underlay innovations in military technology by private firms before World War II, when there was no recognized mechanism for issuing R&D contracts.)

At the time TI and Fairchild began producing the first ICs, selection pressures favored chips that were small, rugged, and reliable. Price was nearly immaterial, since the new devices were intended for NASA missions into space and the guidance systems of intercontinental ballistic missiles. Even so, costs came down rapidly as the two firms and others that joined them moved down their learning curves. To open the way for higher-volume commercial applications, IC manufacturers began to cut prices in advance of realized cost declines, seeking to establish dominant market positions. Industrial sales grew, outstripping government sales, and selection pressures shifted. Metal-oxide semiconductor (MOS) ICs split off from bipolar chips to form a fast-growing new branch on the family tree. (MOS devices are unipolar, utilizing electrical charges of only one sign; bipolar chips make use of both positive and negative charges.) Slower but denser and in part for that reason less expensive on a price-performance basis (more transistors can be crammed onto each chip to offer greater functional capability), sales of MOS devices grew rapidly. Sales increased still faster as prices fell to the point that MOS chips could be designed into mass-market consumer products such as digital watches. The military, meanwhile, continued to demand the highest possible speeds for applications such as signal processing. Since government purchases were declining relative to commercial sales, the appeal of designing chips for the military diminished. By the end of the 1970s, military sales accounted for less than 10 percent of the IC market and DOD was falling noticeably behind the pace of innovation in commercial electronic systems.

Then as now, semiconductor firms competed fiercely. The advantages they gained from innovations in chip design and fabrication were often fleeting but nonetheless vital. The selection environment was harsh. Rapidly increasing capital intensity compelled firms, many of them relatively small and with limited sources of financing, to invest large sums in manufacturing facilities. Over the 1970s, the costs of a new plant rose from perhaps $5 million to around $50 million (on the way to billions) and the money had to be committed well in advance of revenues from the chips the plant would produce. Semiconductor manufacturers thus competed for access to capital too, and to hire the most creative circuit designers and process engineers. Those that fell even slightly behind risked entering a downward spiral that might end in merger with a stronger rival or purchase by a foreign-based multinational.

For a time, selection mechanisms in defense and in commercial markets overlapped somewhat. Both computer manufacturers and the military, for example, wanted high-speed chips for digital processing units and high-speed memory for interfacing with those processing units. But the military also demanded features such as radiation hardness—resistance to the ionizing radiation produced by nuclear blasts—that was of no interest commercially. As DOD lost the ability to influence technological direction, by the end of the 1970s only a few niche suppliers catered to the Pentagon's needs and the Office of the Secretary of Defense (OSD) put in place the Very High-Speed Integrated Circuit (VHSIC) program, which eventually spent nearly $1 billion in an effort to exploit innovation on the commercial side of the industry.

VHSIC was strictly an R&D program.[17] DOD did not attempt to select technical directions or steer commercialization as the Atomic Energy Commission had earlier done with nuclear power. In that case, government itself selected the technology, making little attempt to explore alternative reactor designs. Rather than fostering variety as a starting point for selection, government proffered subsidies for light water reactors based on designs for the Navy. For Rickover and the naval reactors group, the overriding selection criteria had been compactness, along with safety and reliability. The latter would be achieved through highly trained and closely supervised operators and technicians, something the Navy knew how to provide. Utilities had little choice but to implement active safety procedures too, likewise reliant on human operators. Yet they drew on a different labor pool and had no way to command military discipline, which meant that even the best-run utilities could not, in the early years, approach the reliability levels routinely attained in naval service.

For jet engines and gas turbines, military and commercial customers assigned high priorities to similar, though not necessarily identical, performance attributes. From the beginning, DOD wanted jet engines for fighters with excellent transient response during violent maneuvers, which was irrelevant to airlines. But military and commercial customers alike have always valued fuel efficiency. For airlines, saving a fraction of a cent per seat-mile on jet fuel can make the difference between profit and loss. For the military, greater efficiency means longer range for bombers and greater combat radius for fighters. With this selection criterion in common, military R&D and procurement benefited commercial customers quite directly.[18] Indeed, jet engines sold in the civil aviation market have generally been derivatives of military engines (this had also been the case for piston engines before World War II).

Once managers of firms competing in commercial markets have made their decisions and implemented them, sorting of innovations takes place impersonally and automatically, unlike in defense where powerful figures in the armed forces, OSD, and Congress struggle to impose their wills. Selection in defense resembles what we might observe if, say, General Motors (GM) made all technical decisions concerning automobile design. That would not mean decisions imposed from the top. Different groups, divisions, and executives within GM compete for resources; like other big corporations, GM is no more a unitary organization than is DOD. At the same time, both have established hierarchies of authority, which means that the narrowing, pruning, and trimming of technological variety proceed far more deliberately than when independent organizations compete to innovate—the actual case in the global automobile industry or the semiconductor industry.

Selection internal to an organization depends in part on the judgment and preferences of individuals in positions of influence and authority, functioning something like the intuition of the scientist, and in part on nominally rational decision processes looking ahead to an uncertain future. Following the Korean War, the primary selection principle for the U.S. military was straightforward: government would support almost anything and everything that might lead to military advantage vis-à-vis the Soviet Union. The search for technological superiority began with the generation of many alternatives. Some of these were pursued at least through the stage of exploratory R&D, with the most promising (or those with the most political support) approved for prototype construction and testing. At various points, those that did not meet expectations were "selected out." Government spent around $1 billion on both Dyna-Soar and the nuclear-powered bomber (totals that would be much higher in today's dollars) before the programs were cancelled. Neither reached the demonstration stage.

This approach left the services with more alternative weapons systems than the United States arguably needed, including, for instance, six different ballistic missiles (three intercontinental, Atlas, Titan, and Minuteman, and three of shorter range, Jupiter, Thor, and Polaris) and the long list of supersonic fighters purchased in the 1950s. If one of the new systems did not work—or did not fit whatever strategy and doctrine the responsible service eventually adopted—another might. Costly and messy as it was, elements of this approach still survive, as illustrated by DOD's many antiarmor weapons. In the later years of the Cold War, the Soviet Union fielded more than 50,000 tanks. "The technical sophistication and the sheer numbers of these armored vehicles were far greater than the armored threat associated with any other war-fighting contingency, either then or now."[19]

With the collapse of the Russian military, the numbers of tanks that U.S. forces might conceivably face dropped by over four-fifths. Nonetheless, the services had 15 new antiarmor weapons under development or in production in the late 1990s, scheduled to join the 40 antiarmor weapons already in their inventories.

Overlap or duplication in capabilities provides insurance against technical failure or intolerable cost overruns, and a broad portfolio of weapons confers tactical flexibility, provided the weapons work well and are available when and where needed. But the United States has paid a high price for returns that in some cases seem diminishingly small.

Diverging Pathways

For over 50 years, despite continuing European excellence in technology and science and Japan's challenge during the 1980s, the U.S. national system of innovation has been unrivaled. It has demonstrated advantages in both generation of variety and selection. After World War II the federal government began to channel funds to R&D, doing so, as is usual in the United States, in decentralized fashion. Ten agencies now enjoy annual R&D budgets of $500 million or more; over a dozen others get smaller sums. Sometimes criticized for duplication and waste—17 agencies support over 700 federal laboratories, so many that it took a census by the General Accounting Office to determine the total—decentralization has been a major source of flexibility and dynamism. R&D dollars flow in a greater variety of channels than in other national innovation systems. The United States has many more research universities, and more venture capitalists and entrepreneurial firms. Dense networks of formal and informal ties linking research groups with sources of funds and entrepreneurs with sources of ideas foster cross-fertilization of ideas. High labor mobility enables firms to bring in knowledge from outside by hiring experienced engineers and managers, including those poached from competitors, as well as students fresh from graduate programs in research universities.

The selection environment in the United States has also been more rigorous than elsewhere. The private-sector economy, never as tightly regulated as in most other industrialized nations, has become less concentrated and more entrepreneurial over recent decades. Since 1980, for example, the small-firm share of business-funded R&D has quadrupled.[20] Deregulation in industries ranging from trucking to telecommunications has strengthened competition and thus selection pressures, one reason why the United States attracts so much inward investment by foreign-based companies that believe they must compete in U.S. markets to keep pace in innovation.

Defense shares in the first of these sources of strength, but not the second. The services and DOD-wide agencies such as DARPA maintain R&D portfolios diversified by discipline, by objective, and by performer (the military's own laboratories, defense contractors, universities, and other nonprofits). In aggregate, these portfolios are both broad and deep, in part reflecting the long-standing policy of hedging against technological surprise. A host of different subagencies such as AFOSR enjoy considerable freedom in setting research directions. For more applied work, however, DOD has no basis for making decisions comparable to that of businesses when they allocate funds based on anticipation of future profits. This means that well-placed internal advocates can sometimes push through projects without the rigorous screening common in well-managed firms. Once approved, these projects may persist indefinitely:

> If, as I discovered as the deputy chief of naval operations in charge of analysis, planning, and programming on the Navy Staff, ice-penetrating sonar buoys were "out" because of the collapse of the Soviet Union's submarine force, then the same project could continue, perhaps with even greater funding, by calling it the "in"—very "in"— littoral warfare sonar buoy project.[21]

As military technologies such as those related to supersonic flight and stealth, and the very different technologies of digital electronics, diverged from technologies of interest to commercial customers and as the manufacturing sector of the nation's economy, in which formal R&D remains a prominent source of innovation, peaked and began to decline as a share of gross domestic product (GDP), spin-off potential also peaked and began to decline. The manufacturing share of GDP has fallen by more than half since the 1950s, while the service sector, which at the time of World War II already accounted for half of national output, continues its steady expansion. Although many service firms are highly innovative, they do not innovate in the same ways as manufacturing firms (in part because so many of their products have no predetermined identity). Because service firms do not conduct nearly as much formal, budgeted R&D, many of their innovative activities go unmeasured, unreported, and to some extent unrecognized.

Since the 1980s, the nation's military and commercial innovation systems have, for practical purposes, been separate and distinct. They cannot be reintegrated. The period of close coupling, overlap, and symbiosis was brief, an artifact of the Cold War, while the differences between military and commercial innovation are deep and permanent. Some connections can be rebuilt or strengthened through policies such as dual use. More effective implementation of these policies would provide greater U.S. national security at more reasonable cost. But it would have no effect on military requirements and make little difference for the overall design of military systems. That is where the problems of acquisition begin and where we must look for solutions.

CHAPTER 11

TAKING REFORM SERIOUSLY

[I]nstead of policy determining strategy, and strategy in turn determining its military implementation. ... the unilateral aims and policies of the military services are combining to make the strategy they are supposed to serve, and the strategy is tending to make the national policy.

—Commission on Organization of the Executive
Branch of the Government, 1949[1]

The Cold War presented Washington with a designated adversary, one that posed closely studied threats against which new weapons could be gauged. Today, the United States lacks such gauges. Looking ahead, no one can say where U.S. forces might be asked to fight, with what objectives, under what sorts of politically imposed constraints. Congress and the administration of President George W. Bush have given the Department of Defense (DOD) generous budget increases, but the services will always want more and policymakers will always have to decide which of the weapons systems put forward by advocates merit a claim on funds. Decision-makers have less of a basis for making those choices today than during the Cold War.

The institutional settings for choosing weapons originate as far back as World War I. They took on their current forms mostly during the 1950s. Many reforms have been proposed since and some have been implemented, but the institutions themselves have changed relatively little. They have never functioned very well. The core problem is rivalry within and among the services. To be effective, reforms must temper both intra- and interservice competition and at the same time strengthen incentives for buying what the armed forces need collectively to fulfill the missions the nation assigns them rather than what they want individually.

Broad accord on choice of weapons depends on reasonable agreement concerning overall security policy and strategy, which is fundamentally a civilian responsibility, and, following from that, agreement on the military systems and equipment best suited to implement future tasks, which is a joint responsibility of military and civilian officials weighted heavily toward professional military expertise. Accomplishing this will require strengthening both the Office of the Secretary of Defense (OSD) and the Joint Chiefs of Staff (JCS), and in particular the powers of the JCS chair. The Chiefs have not been able put forward a coherent view of military needs because the members (and the joint staff) argue for the positions of the services they represent and the chair has no leverage to override them. Since those service positions are often themselves incoherent—according to an Air Force general, "We have always [programmed] what the Air Force will buy. . . . [without] a description of how it will fit together and fight"—the budget requests of the services taken together can hardly be more so.[2] Even if the chair were able to set aside the claims of his own home service, the powers of the chair are inadequate for brokering agreement on a unified program. The Secretary of Defense does have the authority, in principle. In practice, however, the services have often been able to wear down or outwait the secretary and his staff. OSD, then, should also be strengthened. There are two reasons. First, to counterbalance the increased authority of the JCS and its chair, preserving the current civil-military balance. Second, to enable Pentagon civilians (and civilian officials elsewhere in the executive branch, e.g., in the Office of Management and Budget [OMB]), to resist Congress, when, for instance, service advocates appeal directly to powerful members or committees. In short, the influence of the services over acquisition must be weakened vis-à-vis the JCS and its chair and the influence of civilian officials strengthened.

This final chapter begins by examining the two most commonly suggested directions for acquisition reform, more competition and better management, finding them to be useful but inadequate. Because no functioning market exists in defense and because production volumes are small, there is no solution in economic competition, desirable as this may be when feasible. Given the incentives that operate in defense, managers in both government and the defense industry see their fundamental task as keeping the money flowing. They routinely promise more than they can deliver, overstating capabilities and understating costs. If they did not, the money would flow elsewhere. Firms in the civilian economy put in place counterincentives to curb exaggeration, deception, and dissembling by managers. They base decisions, as best they can, on transparent assumptions and reasoning. Market dynamics reinforce those incentives by exposing bad ideas and flawed innovations that have passed internal screens. None of this exists in defense. Because managerial hierarchies are weak and the true tests of weapons come only in war, advocates can make extravagant claims with little fear of being held to account.

Strengthening managerial hierarchies inside DOD is the only pathway to meaningful reform of major programs. Greater centralization of power and authority runs counter to U.S. political traditions. Yet there is no alternative. If some other route to reform existed, it would long since have been discovered and pursued.

More Competition and Better Management

In 1991, the Army began a new program, planned to run until 2022, under which it expected to buy 85,500 medium trucks at a total cost of some $16 billion. A firm with no experience in building trucks, Stewart & Stevenson Services (SSS), won the initial procurement contract, bidding a design prepared by Steyr-Daimler-Puch, an Austrian company that had been in the truck business for decades. After SSS won the contract with a top-rated design and low-cost production bid, the firm ended its relationship with Steyr-Daimler-Puch, purchased and retooled a plant in Texas formerly used for making oil well equipment, and began turning out vehicles that consistently failed the Army's acceptance tests.[3] This could be a story from World War I or the 1920s.

When the U.S. government sets out to buy military systems and equipment, it faces a fundamental dilemma: How to get what it needs at a reasonable price in the absence of the discipline imposed by markets? For decades, proposals for reform have circled around the twin poles of more competition and better management. These efforts have accomplished relatively little. The Army could not buy the trucks it needed from those available in the marketplace. No one else needs such vehicles. When the Army proposed a second-source contract to introduce competition, the General Accounting Office (GAO) reported that "The Army does not know whether its plan will reduce costs. It did not perform an analysis to determine whether the added costs . . . would be offset by cost savings. Also, it did not compare the costs and benefits of its plan with those of other program alternatives . . . ."[4]

Competition in defense does exist; but it is seldom market-based (table 9). Some years ago, redesign of the Army Jeep resulted in a vehicle 2 inches too wide to fit two-abreast inside the Air Force C-141 cargo plane.[5] No commercial firm would have made such a decision. If it had, market forces would have weeded out the product, and perhaps eventually the firm itself.

Table 9 Competition in Defense

Mode	Intensity	Examples	Impacts
Interservice	Relatively strong.	Long-range land-based bombers (Air Force) vs. aircraft carriers (Navy).	Moderately effective in selecting for militarily effective systems.
Intraservice	Can be vigorous, depending on relative strength of rivals.	F-22 vs. B-2; aircraft carriers vs. submarines.	Political–bureaucratic competition often overrides military–technological competition.
RDT&E	Strong in research.	Potential competitors for DOD research contracts include universities, federally funded research and development centers, service laboratories, and private companies.	Effective in selecting among research avenues and ideas.
	Weak, with some exceptions, in design and development.	Design competition between Lockheed-Martin and Boeing for the F-35 Joint Strike Fighter is the primary recent example of	When affordable, can help select the superior alternative. (Competition among Soviet design bureaus for aircraft, missiles,

		design competition extending to prototypes. Past competition among jet engine manufacturers is widely credited with improving performance and reliability.	and submarines was probably at least as intense as design competition in the United States.)
Procurement	Weak, unless production volumes are quite large (e.g., for expendables such as munitions).	Positive: Sidewinder missile. Negative: Virginia-class submarine.	In the Virginia submarine example, loss of scale and learning economies outweighed competition-driven efficiency gains.
International	Weak.	F-16, F-35 sometimes compete with European planes such as Rafale (France), Eurofighter Typhoon (multination) for third-country procurements. No serious technological competition between the United States and potential adversaries.	Recently inconsequential, so far as the United States is concerned.

With two sets of firms designing prototypes for the F-35 Joint Strike Fighter, DOD increased the likelihood of getting a good airplane. The choice reflected the large size of the planned procurement, initially expected to approach 3,000 units. And even so there were no plans for competition in production: once Lockheed Martin won the design competition, it became the prime contractor. More commonly, design competitions take limited and artificial forms, restricted to the virtual domains of "paper studies" and analytical modeling.

Since the 1950s, competitive bidding for production contracts has seldom been practical except for high-volume expendables such as munitions, given that savings depend on two (or more) firms each attaining something approaching minimum efficient scale (which normally permits substantial learning economies too). For complex systems produced in small numbers, such as the Navy's Virginia-class submarine (chapter 5), a second source almost always means a higher price tag. Even at the height of the Cold War, moreover, production economies were often sacrificed because procurements were spread over many years to limit annual outlays.[6] By themselves, annual rather than multiyear budget appropriations impose a cost penalty that has been put at about 5 percent (contractors fearing future production cuts, for example, may decline to make cost-saving investments in plant and equipment).[7]

When Boeing designs a new commercial plane, it consults with major airlines, each of which has somewhat different needs. Boeing's objective is to satisfy as many potential purchasers as possible. The company pays the bills for design and development, betting that airlines will buy the resulting product. When they place their orders, airlines tell Boeing how many seats to install, where to place the galleys and lavatories, which engines they want (if there is a choice). Boeing makes the decisions on range, takeoff weight and payload, cockpit design, cruising speed and flight characteristics. Defense contractors do not make analogous decisions. When Boeing designs a military airplane, it works from requirements prepared by DOD. Defense firms are hardly bystanders in the process of determining those requirements. They lobby the services, OSD, and Congress, join or support coalitions that back programs they believe will benefit their company, and sometimes initiate those coalitions. Still, the military in the end sets requirements (civilian officials have nominal approval authority but in practice the process is largely controlled by the responsible service) and defense firms must live with them.

Most of the Pentagon's 70,000 or so acquisition personnel deal with small contracts and routine matters; 99 percent of DOD contracts are valued at less than $10 million. Training is "narrow and inadequate. The Defense

Acquisition University largely focuses on contracting skills and does not give enough attention to broad acquisition and program management challenges." It graduates "[u]nder-trained and risk adverse contracting personnel [who] naturally revert to bureaucratic behavior, 'the letter, not the spirit of the regulations.'"[8] And while a military officer who escapes an acquisition assignment without attracting unfavorable notice can look forward to further promotions, a civilian with the skills and experience to oversee a major program could earn more and expect to reach higher management levels in the private sector. In dealing with their counterparts in the defense industry, DOD program offices begin at a distinct disadvantage.

The official Navy inquiry into the aborted A-12 found that

> The A-12 Program has been treated as the Navy's number one aviation priority in fact as well as rhetoric. The PM [Program Manager] is an Aviation Engineering Officer, with three advanced degrees and a career path which would be a model in any of the new Service Acquisition Corps. He has been on-station for more than four years . . . . In short, the PM in this case is the archetype of the well-trained, highly motivated professional, fully empowered to fulfill his responsibility and be accountable for cost, schedule, and performance . . . . Nonetheless, it should be plain that neither he, nor the similarly well-qualified and dedicated officers in his chain of supervision, met the needs of senior civilian leaders within the [Department of the Navy] and DOD for an accurate assessment of the program's status and risk.
>
> [P]rogram-focused managers do not have positive incentives to display the full range of risk in their programs to officials who might respond by cutting their resources.
>
> There is no reason to believe that the factors . . . are unique to this Military Department. Indeed, experience suggests that they are not. Unless means can be found to solve this abiding cultural problem, the failures evidenced in this report can be anticipated to occur again in the same or a similar form. . . . [T]he fundamental problem is to create appropriate incentives to enable senior leaders to rely upon responsible, accountable line managers for realistic perspectives on the cost, schedule and technical status of their programs.[9]

These are familiar themes to students of bureaucracy; in government, it is easy to call for the "right" incentives but hard to put them in place and keep them there. DOD needs a better trained and rewarded acquisition workforce. But in the absence of reforms that alter the way major weapons systems are planned, approved, and budgeted, better program management will not make much difference. No organization can rely on exceptional people to get results in otherwise dysfunctional settings. DOD programs will not be consistently well run until the federal government puts in place a management system that reliably rewards good decisions and penalizes poor decisions, at the program level and above. That is what successful businesses do.

Strengthening Civilian Control and Empowering the Joint Chiefs of Staff

The services base their claims for acquisition dollars on the promise, embodied in requirements, that the systems they want will enable them to accomplish vital military missions. Too often, dissembling and distortion feature in the construction of mission statements and the requirements derived from them. The services, and contractors beholden to them, systematically exaggerate the capabilities of the weapons systems they favor, understate their costs, and denigrate rival systems. In an anonymous interview, an "Air Force civilian official" explained that "The Air Force lies; Congress knows we lie. If everyone told the truth, the one liar would have the advantage." The same informant could not resist adding, "The Navy has some really proficient liars."[10]

Upper-level managers in private firms set the rules for those lower down. They put in place incentives and disincentives to reward honest plans and proposals and penalize deception. DOD managers, by contrast, confront deeply perverse incentives. Better decisions within DOD require better alignment between the structure of incentives and the objectives of national security. That in turn requires strengthened managerial hierarchies on both the military and civilian sides of the Pentagon.

As Secretary of Defense in the 1960s, Robert McNamara was able to increase civilian control over system design and development, which until then had been left almost entirely to the services. Military leaders fought McNamara's policies. Ever since, the troubled TFX/F-111 program has been trotted out to tar the reforms McNamara tried to institutionalize, treated as a proof of failure rather than a lesson to be built upon. With McNamara's managerial reforms further tarnished by his part in the misadventure of Vietnam, the services were able to reassert their control over acquisition following his exit from government in 1968. The consequences included weapons systems such as the B-1B bomber, just the sort of program McNamara and those he brought into OSD had opposed so strenuously.

Since McNamara's departure, the United States has made many changes in acquisition policies and practices. These have had positive effects on routine procurements but have left major programs essentially unchanged. A 2005 *Business Week* editorial points to the reasons. The editorial declared that reforms should "shield the Pentagon from its own specification mistakes or technology excesses." The diagnosis is accurate; yet it is followed by nothing more than a prescription to "more closely link the interests of contractors and taxpayers, perhaps by awarding better bonuses or higher profit margins to contractors."[11] That is the sort of tinkering that so often in the past has accomplished so little. The only way to avoid "specification mistakes" and "technology excesses" is to sharply reduce the influence of the armed forces over requirements. Responsibility should be shifted to civilian officials, who could then ensure that defense firms have more freedom to exercise discretion in managing technological tradeoffs. By issuing broad guidelines in place of the often arbitrary and unstable requirements of the past, DOD would increase incentives for contractors to assign their most qualified engineers and scientists to its programs, and allow them to exercise their creativity and technical judgment. The services would oppose such changes, as they did McNamara's policies. They would have to be pried away from the control they are accustomed to exercising over requirements and their propensity for making end runs to Congress. To accomplish this will require stronger organizational hierarchies on the military as well as the civilian side of the Pentagon.

The overriding objective must be to reduce the ability of any one service to hold out for programs that it alone values. That is best accomplished by strengthening both OSD and the JCS, in particular its chairperson, relative to the services individually. Perhaps the most formidable obstacle to a stronger JCS lies in the suspicions held by so many Americans of concentrated military power. The best way to counter those fears is to strengthen the civilian hierarchy of DOD in parallel, so the civil-military balance does not change.

To function effectively, any such realignment must ensure that civilian officials receive timely and reliable recommendations and advice from military professionals, especially from the JCS, since it alone can craft an integrated military perspective. In the past, the Chiefs have forwarded dilatory compromises that result from the playing out of inter- and intraservice interests. Selecting weapons systems needed for national security has come second to buying what the services want for parochial reasons.

After World War II, General George C. Marshall expressed a widely held view when he said, "lack of real unity . . . handicapped the successful conduct of the war."[12] Notwithstanding world conflict on a scale that tested the will and resources of the entire nation, cooperation between the Army and the Navy was in large measure an achievement of the consummate political tactician who occupied the White House, Franklin D. Roosevelt, a man whose own style and methods combined with improvised arrangements for military planning—improvisation that the President scarcely discouraged—to create more than occasional disarray.

The intended corrective, the National Security Act of 1947, included among its provisions legislative authorization for the Joint Chiefs of Staff, which had been established on an informal basis during the war to coordinate decision-making by the service chiefs. From those beginnings, it has continued to function as a weak committee with a weak chair. Indeed, Congress did not give the Chiefs a chairperson for another two years, and the chair had no vote until 1958.

Soon after the 1947 National Security Act took effect, some 120 interservice boards and committees were busy with inconclusive debates that gave the appearance of coordination while leaving the services free to do largely as they wished. The Chiefs too have been happy to occupy themselves with minor issues on which they could be seen to disagree so long as those issues did not threaten core service interests. At the same time, the Chiefs avoided most big decisions. The services wanted it that way and so did politicians, aware of the fears held by so many Americans of concentrated military power, or indeed concentrated government power of any sort. Already in 1949 the Hoover Commission was calling for further organizational changes to remedy "continued disharmony and lack of unified planning."[13]

On issues the services care about, the Chiefs count on unanimity to get their way with the rest of government and usually manage to present a common front. Because each service has an effective veto, exercised by the threat to withhold approval and shred the appearance of unity that all rely upon, internal negotiations result in compromises that leave each service free to press its claims with OSD, the White House, and Congress—more ships for the Navy, more planes for the Air Force, more armored vehicles and helicopters for the Army and Marine Corps. Neither the chair nor the JCS collectively have been willing to establish binding priorities. The chairman, even if he had the will (and could put aside ways of thinking absorbed during a lengthy career in which advancement in rank depended on at least the appearance of conformity with the norms of his home service), does not have the power to resolve disagreements. What normally emerges is "the lowest common denominator of what the services can agree on."[14]

These dynamics have not changed much since the early postwar years. In 1948, President Harry S. Truman directed his Secretary of Defense to return a fiscal 1950 budget totaling no more than $14.4 billion. This was the first effort to prepare a unified defense budget in place of the independent Army and Navy budgets of previous years. The Chiefs responded with service requests adding up to more than twice that sum, setting off a long-running conflict with the Budget Bureau and the president that the Secretary of Defense had no way to resolve. As the dispute dragged on, "the three services were scaled down to the minimum budget [Truman's $14.4 billion] on a roughly equal basis."[15] More than half a century later, that remains a common outcome, posturing to the contrary notwithstanding. Rather than a DOD budget based on an agreed strategic posture, the nation's net military capabilities stem from a series of compromises that keep the service shares relatively constant, as we saw earlier (figure 3 in chapter 3).

Mutual accommodation has consequences for operations as well as acquisition. Although it might seem that the services would be more willing to put aside their differences when fighting a common enemy than when battling over the budget, poor interservice coordination was endemic, as it had been in World War II, during "America's involvement in Vietnam and culminating with the Iran hostage rescue attempt and the Beirut bombing. . . . Even the 1983 Grenada intervention, where the United States *won*, caused serious concern over the lack of progress in executing joint operations."[16]

Grenada was enough of an embarrassment that Congress, in the 1986 Goldwater-Nichols Act, directed the JCS to provide consensus military recommendations to operational commanders as well as the Secretary of Defense and President. The intent was to strengthen what the military calls unity of command. Despite the changes, reports of poor interservice cooperation cropped up during the recent fighting in Afghanistan and Iraq.[17]

Two decades after Goldwater-Nichols, it seems plain that further strengthening of the JCS is needed. The chair should be given more power and the four service chiefs, who normally retire from those positions (unless selected by the president as JCS chair), should be given reasons to moderate their advocacy of service positions in favor of a more collegial perspective. The JCS also needs a larger, better-qualified, longer-serving, and more independent staff. Staff tours averaging under two years have obstructed development of institutional capacity. Promotions controlled by a body independent of the services would encourage retention of capable officers.

Generally speaking, military operations should be left to the military. That cannot be said of acquisition. While OSD, in the person of the Under Secretary of Defense for Acquisition, Technology and Logistics, has formal authority over acquisition, in practice that authority is hard to exercise and many of OSD's bureaucratic resources are consumed by endless wrangling with over marginal issues. So long as effective control of weapons programs remains with the services there will be little meaningful reform. Because poorly formulated or unstable requirements are the starting point for so many acquisition problems, control over requirements must be an explicit part of the transfer of authority to civilian officials.

Any move to strengthen the formal powers of OSD would face objections including the memory of McNamara's efforts to impose civilian authority over acquisition. Some commentators would probably portray this as an unmitigated disaster, and the more heated opponents might conflate the political micromanagement of military operations during the Vietnam War with OSD's efforts to rationalize weapons purchases, a different matter entirely. The conflict between George W. Bush's Defense Secretary, Donald Rumsfeld, and the armed forces, likewise involving both military operations, principally in Iraq, and acquisition, would add fuel to this set of objections. That does not change the fact that DOD civilians are better positioned to exercise unbiased judgment concerning acquisition. The role of the armed forces should be to give honest advice.

As we saw in chapter 4, private firms will always try to take advantage of government; the defense industry must be held to account. But the military must be held to account too. That is a job for civilian officials.

Taking Planning Seriously

Any undertaking resembling that outlined above would call for substantial expenditure of political capital. The armed forces have lived through many past shifts in administration policy and OSD management style. No part of the federal government (except Congress itself) can match the Army, Air Force, Navy, and Marines in time-tested ability to look out for institutional interests. It will take a concerted effort by Congress and more than one administration, pursuing consistent policies over at least a decade, to convince the armed forces that times truly have changed.

It will also take a degree of clarity and consensus on national security policy lacking since the end of the Cold War. These matters have been often pronounced upon without much in the way of reasoned debate. Critics had little difficulty in portraying the Army's Crusader—a 40-ton self-propelled artillery piece that would have been followed into battle by an equally weighty ammunition carrier—as a dinosaur from the age of industrial warfare ill-suited to the age of information warfare. The more difficult questions involve not so much which weapons should drop off the bottom of service wish lists as what to buy from higher up on those lists. How many Stryker combat vehicles for the Army, cargo planes for the Air Force, amphibious assault ships for the Navy? Unmanned combat air vehicles? Hypersonic cruise missiles? Battlefield robots? This book cannot address such questions. But we have already seen why sensible choices so often prove so difficult: the military's normal response will be "all of the above," and what the nation ultimately purchases will then depend not on some more-or-less sober assessment of possible threats but on how much money Congress and the administration end up making available.

Uncertainty concerning future threats tends to evoke a reflexive if sometimes unstated reaction from the military: whatever might come along in the years ahead can be viewed as a "lesser included case" of major theater war.[18] A convenient justification for programs such as the F-22 and part of the visionary impulse behind the Revolution in Military Affairs, this hardly seems a universally valid axiom. The post-invasion fighting in Iraq, in which U.S. forces suffered not from any lack of high-technology systems but from shortages of effective body and vehicle armor, is only the most recent illustration.

Like other outbreaks of irregular conflict since the end of the Cold War, the effort to pacify Iraq demonstrates that the U.S. military is likely to confront a wide range of future tasks, including small- to medium-scale conflicts as in the Balkans and Afghanistan and "operations other than war" such as peacekeeping and counterterrorism, that do not much resemble the wars for which they have prepared since Korea. Military leaders rarely raise the question of whether little wars call for different systems and equipment than big wars. The *National Military Strategy* released by the JCS in 2004 states, "The United States will conduct operations in widely diverse locations [which] may be dramatically different than the high intensity combat missions for which US forces routinely train." Elsewhere the document says, "The goal is Full Spectrum Dominance—the ability to control any situation or defeat any adversary across the range of military operations."[19] Nowhere do the authors ask whether such statements might sometimes conflict. Part of the reason is perhaps that the services prefer to avoid "operations . . . dramatically different than . . . high intensity combat" and content themselves with lip service to preparation for small wars and peacekeeping. The "Chairman's Assessment" appended to the most recent Quadrennial Defense Review (QDR) states that "the QDR recognizes Stability, Security, Transition, and Reconstruction (SSTR) as a U.S. government wide mission of increasing importance and identifies military support to SSTR as a core mission."[20] That does not read like a strong endorsement.

For nearly half a century, the Soviet Union posed the defining threat for the design of U.S. weapons systems. The threat solidified in a period of geopolitical turmoil during which Washington, unaware of the fissures already opening between the USSR and China, viewed itself as confronting a monolithic adversary committed to the worldwide spread of Marxist-Leninist rule. The nation's political leaders gave U.S. military planners two fundamental tasks: to maintain a massive and reliable nuclear deterrent, and to be ready to fight the Red Army in Central Europe. The chief secondary task was to check Soviet adventures and Soviet-inspired adventures in other parts of the world. Year after year, U.S. spy planes and surveillance satellites photographed Soviet air bases, missile silos, and shipyards. The United States and its NATO allies studied possible paths of advance by the tanks and troops of the Warsaw Pact, compiled target lists for NATO's own bombs and missiles, prepared alternative operational plans and tried them out in war games and maneuvers. The Pentagon accepted techno-logical innovation, indeed insisted on it, because in almost any scenario NATO forces would be outnumbered.

Each year the U.S. military pressed for more and better weapons. Planners knew what more and better meant. The Cold War, like all arms races, had its yardsticks. U.S. and NATO postures could be compared to Soviet and Warsaw Pact forces in numbers and capabilities of troops, tanks, and planes. The U.S. government might classify its intelligence estimates, but arbitors such as the Stockholm International Peace Research Institute presented the standings each year for all to see: the arms race could be tracked almost as easily as a baseball pennant race. The U.S. Air Force bought planes designed to penetrate and suppress the air defenses of another superpower, the Army fast M1 Abrams tanks to speed to choke points where their highly accurate fire could disrupt advances by the Red Army. It is no exaggeration to say that U.S. weapons were designed to fight World War III.

Without too much oversimplification, the 1991 Gulf War can be taken as a punctuation mark signifying the end of one era of uncertainty, con-cerning the outcome of a war that never came, and the beginning of another, characterized by quite different unknowns. Until a few months before Saddam Hussein's invasion of Kuwait, after all, the United States had been passing intelligence to Baghdad, a continuation of aid provided during Iraq's earlier war with Iran, considered by U.S. policymakers to be under Soviet influence. It would be another decade before the full extent of the dangers associated with the new era were driven home, with appalling force, on September 11, 2001. In the aftermath, potential threats have come to seem amorphous, the measuring rods elastic. During the Cold War, "readiness" meant "forces in being" to meet Red Army troops pouring through the Fulda Gap. There are no comparable gauges today. Instead, the United States faces terrorists abroad and potentially at home, irregular combat as in Afghanistan, "part-time" fighters who hide among peaceable residents in Iraqi cities, and secretive states such as North Korea, about which U.S. intelligence agencies may know less than they did about Iraq prior to the 2003 invasion.

The awe-inspiring power of U.S. weaponry itself contributes to the new threats. The Pentagon has left potential adversaries with few options other than unconventional or asymmetric forms of fighting, one of which is terrorism. An open society like the United States will always be vulnerable to at least some of these threats. In the absence of an international regime for regulating trade in arms, the nation's enemies, again including better-financed terrorist groups, have little trouble purchasing quite advanced weapons (e.g., shoulder-fired missiles capable of downing civilian airliners). And in the aftermath of the U.S. invasion of Iraq more states may be tempted to seek nuclear weapons, seeing them as a deterrent against overwhelming U.S. conventional power obedient to a government that, from their perspective, may seem to act erratically and sometimes rashly. In a world of fluid politics, fickle alliances, and potential forms of aggression including info-warfare and bio-terrorism, the United States finds itself confronting dangers quite unlike those for which it prepared so assiduously during the second half of the twentieth century.

The U.S. military has made many adjustments in force posture and planning since the withering of Soviet military power and the September 2001 attacks on New York and Washington. Yet there is no reason to suppose that the weapons now under development will prove suited to a future that is bound to be different from that envisioned when systems such as the F-22 were designed. There is nothing like another superpower in sight. The United States is alone in possessing a large and diversified inventory of high-technology weapons and intelligence systems. Russia retains a large army and air force, but cannot afford to train its troops and pilots or maintain their equipment. Other states, including North Korea and China, have large armies; none can call on naval or air power remotely comparable to that of the United States. In tonnage, the U.S. Navy surpasses the navies of the rest of the world combined by a factor of perhaps 2; it has 12 aircraft carriers (1 is to be retired), other nations 5 (3 are British). The United States has more early warning aircraft than the rest of the world (and half those belong to Britain and Japan) and nearly twice as many military satellites.[21]

Threats do not materialize overnight; it would take years for a potential adversary to build large-scale offensive capabilities, more years than it should take the United States to respond, given its unmatched technological capabilities. Any serious buildup would be far from invisible. Submarines, missiles, and air defense radars can be counted, their performance assessed. While a secretive state may be able to keep its nuclear weapons laboratories hidden, if an army is to command the confidence of its political masters, troops must train, conduct maneuvers, practice the arts of war. These activities cannot be disguised for long.

In the future, the U.S. military will more likely be asked to apply metered levels of force, or credibly threaten to do so, than to fight and win a high-intensity conflict against a large, well-trained, and well-led enemy. The services nonetheless continue to push for weapons of ever greater complexity and cost, to the point that GAO, known for being critical but not for overstatement, has begun to call them "megasystems," on the basis that "5 years ago, the top five weapon systems were projected to cost about $291 billion combined; today, the top five weapon systems are projected to cost about $550 billion."[22] Compared to such programs, the $11 billion Crusader, which OSD managed to kill only with effort, seems almost incidental.

It does not appear that the U.S. military has worked very hard to think through plans and doctrine for operations other than major theater war. Small wars and operations other than war come in many varieties. Since World War II, the armed forces have not welcomed them.[23] Despite a lengthy history of occupation and administration, not only in Germany and Japan after 1945, but, among other places, in Cuba and the Philippines after the Spanish-American War, the services have not scoured history, not their own and not that of other militaries, for lessons, tested the possibilities via simulations and exercises, developed a generalized basis from which to plan for particular assignments, whether halting atrocities in Bosnia and Kosovo or pacifying Iraq.

For whatever reasons the leaders of the U.S. military appear to still believe, as they have since Korea, that a large and diverse inventory of high-technology weapons will confer flexibility in meeting whatever future threats may appear and, as a corollary, that a shift in capabilities and doctrine away from major theater war as their primary mission would prove broadly debilitating, harming prospects of prevailing quickly and with few casualties in other sorts of conflicts under other conditions. As a generalization, the first of these claims may be valid, depending on the size of the inventory of weapons and whether it is truly diverse or merely duplicative. The second is dubious, the fallacy exposed in Vietnam, a war for which U.S. doctrine proved inadequate and U.S. weapons ill-suited. Yet even if the premises were accepted, the question still remains: What, exactly, should the nation purchase, given that there is little realistic prospect of buying everything the services want?

In principle, acquisition decisions reflect threat assessments that have been widely analyzed and debated. In practice, such assessments have often been vague and open-ended, or simply ignored in making budgetary requests. That was true even at the height of the Cold War, notwithstanding widespread agreement on the dimensions of the Soviet threat. During Ronald Reagan's first term as president, many billions of dollars went for new weapons without anything resembling a careful evaluation of the nation's defense posture. The services simply brought out lists of systems that had not made the cut in previous budgets—the B1-B is only one example—and the White House allocated the funds. An anonymous Reagan administration OMB official recalled in an interview that "[The White House] just wanted to add dollars, which they did. . . . beyond our wildest imagination. It probably was even beyond the wildest expectations of the defense establishment. . . . There was no substantial new initiative. It was just a lot more of the same."[24]

Congress reacted by mandating, in the 1986 Goldwater-Nichols Act, full-scale defense reviews every four years. When the first of these QDRs appeared, in 1997, it drew criticism for ducking areas of major disagreement among the services and seeming to accept and in places advocate a business-as-usual future. That in fact appears to have been the intent. According to General Ronald Fogleman, Chief of Staff of the Air Force at that time, "An Army two-star from the JCS came by . . . and said, 'I have a message from the chairman [Army General John M. Shalikashvili], and the message is, that in the QDR we want to work hard to try and maintain as close to the status quo as we can.' "[25]

A widely noted statement in that first QDR read:

> [I]t is imperative that the United States now and for the foreseeable future be able to deter and defeat large-scale, cross-border aggression in two distant theaters in overlapping time frames . . . .[26]

Presented as an assertion rather than argued, critics saw in the two-front scenario little more than a pretext for increases in defense spending.[27] The most recent QDR calls for "the capability to conduct multiple, overlapping wars." It also features a good deal of rhetoric concerning

> the non-traditional, asymmetric challenges of this new century [including] irregular warfare (conflicts in which enemy combatants are not regular military forces of nation-states); catastrophic terrorism employing weapons of mass destruction . . . ; and disruptive threats to the United States' ability to maintain its qualitative edge and to project power.[28]

It does not, however, inquire at all deeply into what this might mean for future choices of weapons systems. The document lists "QDR Decisions" for each service, but they are unenlightening. The first two entries on the Navy list, for example, are "Build a larger fleet . . . ." and "Accelerate procurement of Littoral Combat Ships . . . ."[29] How big a fleet? Accelerate by how much? Those questions are to be deferred:

> [DOD] will continue the process of continuous reassessment and improvement with periodic updates in the coming years and by directing the development of follow-on "roadmaps" . . . .
> The full budgetary and programmatic implications of the QDR will be reflected in the upcoming budget cycle.[30]

A generalized emphasis on the open-ended nature of future threats cannot serve as a guide for decision-making. Asymmetric attack has presumably been a staple of tactical cunning since cave dwellers first began fighting one another in groups. The questions, as always, are how, exactly, to maintain flexibility, keep the opponent guessing, operate on faster decision cycles, and otherwise "fight smarter." One might think such questions would be routinely addressed in DOD foresight and planning exercises. To some extent they are. Yet as we have so often seen in earlier chapters, planning, though ongoing, is disconnected from the budgetary decisions which in the end determine the weapons actually built.

During the 1960s, the services lost a portion of their influence over choice of weapons to McNamara's OSD. In the 1980s, they regained the ground they had lost and more. According to Admiral William A. Owens, former vice chairman of the JCS,

> The reassertion of various military service prerogatives reached a post–World War II high during the Reagan Administration from 1981 to 1988. Responding to the administration's eagerness to build up U.S. military power and its belief that the military itself understood its own needs best, each of the individual services was able to control—more than ever before—the defense secretary's formal defense guidance document for that service's own military programs, and the civilian bureaucracy in the office of the secretary of defense retreated from that responsibility.[31]

The civilian staff of OSD peaked in the late 1960s and remains smaller today than under McNamara.[32] According to a recent report of the Defense Science Board, the competence of OSD personnel has also declined: "The staff [of the Under Secretary of Defense for Acquisition, Technology and Logistics and the Director of Defense Research and Engineering] . . . does not come close to matching the breadth and depth of technical talent and experience inside OSD during much of the Cold War."[33] Meanwhile, the services have enhanced their ability to influence policy decisions, in part by sending large numbers of officers to graduate school for degrees in fields including political science and management so they can better uphold service views in intragovernment debates.[34]

To considerable extent, how much money each of the services gets for its programs reflects understandings reached inside the armed forces, hidden at least partially from civilian officials and almost totally from the public at large. Their monopoly over professional expertise gives the services control over assumptions and context. Officers who advance to high rank absorb the traditions, values, and viewpoints of their service and service branch and at the same time acquire credibility outside the military because most Americans believe that their military functions as a meritocracy, which it does, albeit on its own terms. The services have not been above releasing after-action reports in the wake of recent conflicts that "read more like public relations documents than like serious and thoughtful analyses of what happened and why."[35] They have a further advantage in continuity. If one of the services does not get what it wants this year or in this administration, it can simply wait until the next election cycle brings a fresh group of elected and appointed officials.

In view of all this, it may seem too easy to call for a return of planning, budgeting, and overall acquisition authority to civilians. Yet that is the necessary starting point for reform. Defense planners must look ahead to an unknowable future. Who will be the enemy? What will be his capabilities, strategy, and doctrine? What will U.S. political-military objectives be? The armed forces have no monopoly on foresight. Indeed, the historical record shows they are not very good at it. The services must be forced to make the case for what they want, in the open, to OSD and OMB, the White House, Congress, and to the public at large. Transparency must increase, ideas good or bad—missile defense, arsenal ships, nonlethal weapons—get an open hearing. Congress must discipline itself to reduce micromanagement and pork barrel spending and the public must demand an end to the obfuscatory appeals to special knowledge that the services and their acolytes use to push through programs they want. Strengthening civilian authority inside the Pentagon is not a panacea, but it is the only road that will lead to genuine reform. National security can only benefit.

NOTES

Chapter 1   Choosing Weapons

1. The figures in this paragraph are based on *Historical Tables, Budget of the United States Government: Fiscal Year 2008* (Washington, DC: U.S. Government Printing Office, February 2007), Tables 3.2 and 9.7.
2. Statement of Katherine V. Schinasi, Managing Director, Acquisition and Sourcing Management, U.S. Government Accountability Office, "DOD Acquisition Outcomes: A Case for Change," Before the Subcommittee on AirLand, Committee on Armed Services, U.S. Senate, November 15, 2005, GAO-06–257T, p. 2. DOD's plans and projections for weapons programs change constantly and the specifics given in this chapter and those that follow should be viewed in that light.
3. James G. Burton, *The Pentagon Wars: Reformers Challenge the Old Guard* (Annapolis, MD: Naval Institute Press, 1993), p. 29.
4. The quoted phrase, from an anonymous participant in the 2001 review, appears in David A. Fulgum, "QDR Became 'Pabulum' As Decisions Slid," *Aviation Week & Space Technology*, October 8, 2001, pp. 70–71.
5. *Options for the Navy's Future Fleet* (Washington, DC: Congressional Budget Office, May 2006), p. 36.
6. *The National Security Strategy of the United States of America* (Washington, DC: Office of the President, September 2002), p. 29.
7. DOD originally defined major acquisition programs, 85 of which were underway in 2007, as those with total expenditures in excess of $1 billion in 1980 dollars or, for programs limited to R&D, $200 million. Adjusted for inflation, the floor is now $2.19 billion in 2000 dollars or $365 million for R&D. The chapters that follow skirt many issues of nuclear weapons, which are the responsibility of the Department of Energy, a civilian agency. The Energy Department's budget is large in absolute terms, recently exceeding $20 billion, about three-quarters for weapons-related work (and nuclear reactors for submarines and aircraft carriers). Still, the sums are modest compared to DOD acquisition, budgeted at $180 billion in fiscal 2007.

Chapter 2   Technology and Doctrine

1. *Report of the Defense Science Board Task Force on Training Superiority & Training Surprise* (Washington, DC: Office of the Under Secretary of Defense for Acquisition, Technology & Logistics, January 2001), p. 7. The Navy's "top gun" school, created in response to poor performance by its pilots in the Vietnam War, was the first of the CTCs.

2. Jeter A. Isely and Philip A. Crowl, *The U.S. Marines and Amphibious Warfare: Its Theory, and Its Practice in the Pacific* (Princeton, NJ: Princeton University Press, 1951), p. 369.

3. See, e.g., George W. Baer, *One Hundred Years of Sea Power: The U.S. Navy, 1890–1990* (Stanford, CA: Stanford University Press, 1994), pp. 9–33. Mahan's famous *The Influence of Sea Power upon History, 1660–1783* appeared in 1890.

4. David Alan Rosenberg writes, in "The Origins of Overkill: Nuclear Weapons and American Strategy, 1945–1960," *International Security*, Vol. 7, No. 4, Spring 1983, pp. 3–71:

   From 1951 on, General LeMay [SAC commander] did not submit his annually updated Basic War Plans as required for JCS [Joint Chiefs of Staff] review (p. 37).

   [President Eisenhower] never fully understood the gravity of the disputes over nuclear strategy and targeting which raged in the JCS . . . . and did not anticipate the long term consequences of placing the coordination machinery under SAC domination (p. 69).

5. Disdain for civilian officials comes through repeatedly in, for instance, the interviews with high-ranking Navy officers collected in [Captain] E.T. Wooldridge, ed., *Into the Jet Age: Conflict and Change in Naval Aviation, 1945–1975—An Oral History* (Annapolis, MD: Naval Institute Press, 1995). An example, mild of its type, from a chapter based on interviews with Vice Admiral Robert B. Pirie (pp. 72–73):

   Shortly after Mr. McNamara came in as secretary of defense. . . . [t]hey started . . . to tell us how to design and develop the next generation of aircraft. . . . My particular dislike for them and for their approach . . . was that none of them had any significant experience. Just because they'd been hired by the Rand Corporation . . . didn't qualify them to try to tell us how to do our business.

6. For examples from within the Army, see Richard J. Dunn III, "Transformation: Let's Get It Right This Time," *Parameters*, Vol. 31, Spring 2001, pp. 22–28.

7. General Maxwell Taylor, stationed in Berlin during the early part of the Korean War, wrote:

   I had expected a mushroom cloud to rise from the battlefield at any moment. . . . But I found in Washington that there had been cogent military reasons . . . for having withheld atomic weapons. In the first place, we had too few of them. . . . A second reason was that the mountainous terrain of Korea would have limited the effectiveness of these weapons. . . . Finally, it was feared that their employment there might reveal shortcomings which would have diminished their deterrent effect elsewhere.

Maxwell D. Taylor, *Swords and Plowshares* (New York: Norton, 1972), p. 134.

8. George N. Lewis and Theodore A. Postol, "Video Evidence on the Effectiveness of Patriot during the 1991 Gulf War," *Science & Global Security*, Vol. 4, 1993, pp. 1–63.

9. I.B. Holley, Jr., *Ideas and Weapons* (New Haven, CT: Yale University Press, 1953).

10. Hubert C. Johnson, *Breakthrough! Tactics, Technology, and the Search for Victory on the Western Front in World War I* (Novato, CA: Presidio, 1994), pp. 161–179.

11. I.B. Holley, Jr., "Some Concluding Insights," *Case Studies in the Achievement of Air Superiority*, Benjamin Franklin Cooling, ed. (Washington, DC: Center for Air Force History, 1994), pp. 609–625.

12. Stephen Peter Rosen argues, in *Winning the Next War: Innovation and the Modern Military* (Ithaca, NY: Cornell University Press, 1991), that military innovations, technological and organizational, take root and spread only when a substantial number of officers who have had personal experience with the innovation rise to high levels of rank, responsibility, and influence. During peacetime, promotions take years: diffusion is slow, particularly for innovations that entail significant revisions to doctrine. During wars of some length, junior officers advance and field experience, ambiguous though it may be, helps pinpoint the ingredients that lead to success on the battlefield.

13. Although U.S. Army aviators made plans for bombing Germany during World War I in the expectation that her "manufacturing works would be wrecked and the morale of the workmen would be shattered," no such raids had been conducted. Thomas H. Greer, *The Development of Air Doctrine in the Army Air Arm, 1917–1941* (Washington, DC: Office of Air Force History, 1985 [September 1955]), pp. 10–12, quoting from a 1917 document prepared by the Air Service's chief planner.

14. Gene I. Rochlin, Todd R. La Porte, and Karlene H. Roberts, "The Self-Designing High-Reliability Organization: Aircraft Carrier Flight Operations at Sea," *Naval War College Review*, Vol. 40, No. 4, Autumn 1987, pp. 76–90, describes "an uninterrupted process of on-board training and retraining that makes the ship one huge, continuing school for its officers and men" (p. 80).

15. Stephen Biddle, *Military Power: Explaining Victory and Defeat in Modern Battle* (Princeton, NJ: Princeton University Press, 2004), pp. 29–39.

16. Allan R. Millett, "The United States Armed Forces in the Second World War," *Military Effectiveness, Volume III: The Second World War*, Allan R. Millett and Williamson Murray, eds. (Boston: Allen & Unwin, 1988), pp. 45–89.

17. Thomas L. McNaugher with Roger L. Sperry, "Improving Military Coordination: The Goldwater-Nichols Reorganization of the Defense

Department," *Who Makes Public Policy? The Struggle for Control between Congress and the Executive*, Robert S. Gilmour and Alexis A. Halley, eds. (Chatham, NJ: Chatham House, 1994), pp. 219–258; quotations from p. 229.

18. *Defense Science Board 2005 Summer Study on Transformation: A Progress Assessment, Volume II: Supporting Reports* (Washington, DC: Office of the Under Secretary of Defense for Acquisition, Technology and Logistics, April 2006), p. 201.

19. Thomas A. Keaney and Eliot A. Cohen, *Gulf War Air Power Survey: Summary Report* (Washington, DC: U.S. Government Printing Office, 1993), pp. 149, 169.

20. "Joint Warfighting: Attacking Time-Critical Targets," GAO-02-204R, letter to the Honorable Jerry Lewis from James F. Wiggins, Director, Acquisition and Sourcing Management, U.S. General Accounting Office, November 30, 2001, p. 1.

21. *Beyond Goldwater-Nichols: Defense Reform for a New Strategic Era, Phase 1 Report* (Washington, DC: Center for Strategic and International Studies, March 2004), p. 23.

22. Frederic A. Bergerson, *The Army Gets an Air Force: Tactics of Insurgent Bureaucratic Politics* (Baltimore, MD: Johns Hopkins University Press, 1980), especially pp. 70–78.

23. David M. Keithly, "Revamping Close Air Support," *Military Review*, March–April 2000, pp. 14–22.

24. *Defense Acquisitions: Need to Revise Acquisition Strategy to Reduce Risk for Joint Air-to-Surface Standoff Missile*, GAO/NSIAD-00-75 (Washington, DC: U.S. General Accounting Office, April 2000).

25. Robert F. Coulam, *Illusions of Choice: The F-111 and the Problem of Weapons Acquisition Reform* (Princeton, NJ: Princeton University Press, 1977).

Chapter 3   Korea and its Aftermath: The Shift to High Technology

1. *Camp Colt to Desert Storm: The History of U.S. Armored Forces*, George F. Hofmann and Donn A. Starry, eds. (Lexington: University Press of Kentucky, 1999), pp. 217–262; quotation from p. 256.

2. See, e.g., Peter Padfield, *Battleship* (Edinburgh: Birlinn, 2000).

3. Louis Brown, *A Radar History of World War II: Technical and Military Imperatives* (Bristol, UK: Institute of Physics, 1999), p. 29.

4. Clay Blair, Jr., *Silent Victory: The U.S. Submarine War Against Japan* (Philadelphia and New York: Lippincott, 1975), p. 881. Japanese small arms, artillery, and armored vehicles were greatly inferior.

5. Alan S. Milward, *War, Economy and Society, 1935–1945* (Berkeley: University of California Press, 1977), p. 67; Mark Harrison, "Resource Mobilization for World War II: The U.S.A., U.K., U.S.S.R., and

Germany, 1938–1945," *Economic History Review*, 2nd series, Vol. 61, 1988, pp. 171–192.

6. Mark Harrison, "The Economics of World War II: An Overview," *The Economics of World War II: Six Great Powers in International Comparison*, Mark Harrison, ed. (Cambridge: Cambridge University Press, 1998), pp. 1–42; Table 1.7, p. 17.

7. Kenneth W. Condit, *The History of the Joint Chiefs of Staff: The Joint Chiefs of Staff and National Policy, Volume II, 1947–1949* (Wilmington, DE: Michael Glazier, 1979), p. 191.

8. These are the totals for 1939–1944 given in R.J. Overy, *The Air War, 1939–1945* (New York: Stein and Day, 1981), p. 150.

9. Public statements by U.S. officials consistently exaggerated the size and potency of the Red Army, which was large but not overwhelmingly so and poorly equipped for some time after World War II. John S. Duffield, "The Soviet Military Threat to Western Europe: US Estimates in the 1950s and 1960s," *Journal of Strategic Studies*, Vol. 15, 1992, pp. 208–227.

10. The August 1953 Soviet demonstration released less energy than existing fission bombs; not until late 1955 did the Soviet Union test a hydrogen warhead comparable in power to those in the U.S. stockpile. Richard Rhodes, *The Making of the Atomic Bomb* (New York: Simon & Schuster, 1986), p. 778.

11. Quoted on p. 248 of David Alan Rosenberg, "American Postwar Air Doctrine and Organization: The Navy Experience," *Air Power and Warfare: The Proceedings of the 8th Military History Symposium, United States Air Force Academy, 18–20 October 1978*, Alfred F. Hurley and Robert C. Ehrhart, eds. (Washington, DC: U.S. Government Printing Office, 1979), pp. 245–278.

12. The *Time* quotation appears in Walter S. Poole, *The History of the Joint Chiefs of Staff: The Joint Chiefs of Staff and National Policy, Volume IV, 1950–1952* (Wilmington, DE: Michael Glazier, 1980), pp. 66–67. National Security Council Memorandum 68, a prominent signpost marking the turn towards a stiffer Cold War posture, had gone to the White House in April 1950, well before North Korea's invasion of the south. President Truman accepted NSC 68 in late September. Melvyn P. Leffler writes, in *A Preponderance of Power: National Security, the Truman Administration, and the Cold War* (Stanford, CA: Stanford University Press, 1992), p. 356, "What was new . . . was that Nitze [Paul H. Nitze, the principal author of NSC 68, who was the State Department's Director of Policy Planning at the time] simply called for more, more, and more money . . . . to build up U.S. military capabilities . . . ." Leffler's point is that NSC 68 was in other respects little more than a restatement of existing policies and programs, although some of these were relatively recent.

13. On the change in perceptions and attitude in Washington toward R&D as a result of the Korean War, see William A. Blanpied, ed., *Impacts of the Early Cold War on the Formulation of U.S. Science Policy: Selected Memoranda*

of William T. Golden, October 1950–April 1951 (Washington, DC: American Association for the Advancement of Science, 1995).

14. The poison gases that spread terror on World War I battlefields were offshoots of research on synthetic dyes for the chemical industry, work conducted by the same German chemical firms that earlier had pioneered in industrial research. Ulrich Trumpener, "The Road to Ypres: The Beginnings of Gas Warfare in World War I," *Journal of Modern History*, Vol. 47, 1975, pp. 460–480.

15. George Perazich and Philip M. Field, "Industrial Research and Changing Technology," Report No. M-4, Work Projects Administration, National Research Project, Philadelphia, January 1940.

16. George Wise writes, in *Willis R. Whitney, General Electric, and the Origins of U.S. Industrial Research* (New York: Columbia University Press, 1985), p. 215:

    The exact contribution of industrial research may have been misunderstood. Industrial scientists did not invent light bulbs, long distance telephony, cellophane, radio, or the refrigerator. But they did invent the crucial improvements, such as a ductile tungsten process, wave filters, waterproof cellophane, better vacuum tube amplifiers, and freon refrigerators. These key improvements became the basis for commercial dominance.

17. Before World War I, each of the major European powers had built air armadas of some size—an estimated 500 aircraft for Germany, Russia, and France (generally considered the technical leader), and 250 for Britain. At the time, the U.S. Army and Navy could count 19 planes between them. Archibald D. Turnbull and Clifford L. Lord, *History of United States Naval Aviation* (New Haven, CT: Yale University Press, 1949), p. 40.

18. Over the course of NACA's existence, 1915–1958, military officers and civilian employees of defense agencies accounted for over half (67 of 119) of those serving on the main committee. Alex Roland, *Model Research: The National Advisory Committee for Aeronautics, 1915–1958*, NASA SP-4103, Vol. 2 (Washington, DC: National Aeronautics and Space Administration, 1985), Table B-1, pp. 427–430.

19. James Gregory McGivern, *First Hundred Years of Engineering Education in the United States (1807–1907)* (Spokane, WA: Gonzaga University Press, 1960), pp. 32–39 and 88. Before the Civil War, West Point admitted many more men than the Army needed. Some planned from the outset to pursue careers as "civil" engineers; others sought such work after failing to win one of the few commissions available to new graduates.

20. Bernard Brodie, "Technological Change, Strategic Doctrine, and Political Outcomes," *Historical Dimensions of National Security Problems*, Klaus Knorr, ed. (Lawrence: University Press of Kansas, 1976), pp. 263–306; quotation from p. 274.

21. Sims, soon to be an admiral, was a principal architect of the fleet built by the United States before World War I. The quotation appears in

Robert L. O'Connell, *Sacred Vessels: The Cult of the Battleship and the Rise of the U.S. Navy* (Boulder, CO: Westview, 1991), p. 122. More comfortable with machinery than the army, it was the British Admiralty in 1915, after earlier work on armored cars, that established a "Landships Committee" to explore design concepts for tanks. J.P. Harris, *Men, Ideas and Tanks: British Military Thought and Armoured Forces, 1903–1939* (Manchester, UK: Manchester University Press, 1995), pp. 9–39.

22. William M. McBride, *Technological Change and the United States Navy, 1865–1945* (Baltimore, MD: Johns Hopkins University Press, 2000), pp. 12–35.

23. L. Ferreiro, "Genius and Engineering: The Naval Constructors of France, Great Britain, and the United States," *Naval Engineers Journal*, Vol. 110, September 1998, pp. 99–130.

24. Sealed bids could "be used to advantage only when production costs were well known. [They were] completely useless in purchasing non-standard items. . . ." Robert H. Connery, *The Navy and the Industrial Mobilization in World War II* (Princeton, NJ: Princeton University Press, 1951), p. 202. Otherwise colluding firms would be able, for instance, to rotate in submitting winning bids priced to permit abnormal profits. In part for such reasons, the Navy built airplanes in small numbers at facilities adjacent to the Philadelphia Navy Yard from World War I into the 1950s. William F. Trimble, *Wings for the Navy: A History of the Naval Aircraft Factory, 1917–1956* (Annapolis, MD: Naval Institute Press, 1990).

25. William H. Hallahan, *Misfire: The History of How America's Small Arms Have Failed Our Military* (New York: Scribner's, 1994).

26. Paul A.C. Koistinen, *Arsenal of World War II: The Political Economy of American Warfare, 1940–1945* (Lawrence: University Press of Kansas, 2004), writes, p. 106:

> [T]he bureaus functioned virtually free of direction . . . . They asserted that Congress made appropriations to them, precluding the secretary from transferring funds within the department. Furthermore, the secretary of the navy could create no new offices in the department without congressional authorization. With backing form the Judge Advocate General and close ties with Congress, the bureaus could usually checkmate any secretary challenging their entrenched power.

27. "During the summer of 1940 [civilian technical experts] toured the field and learned about fire control [and] radar. Like the fire control group, the radar group realized that the army and navy were unaware of each other's work." David A. Mindell, *Between Human and Machine: Feedback, Control, and Computing before Cybernetics* (Baltimore, MD: Johns Hopkins University Press, 2002), p. 244.

28. E.g., Kurt Hackemer, *The U.S. Navy and the Origins of the Military-Industrial Complex, 1847–1883* (Annapolis, MD: Naval Institute Press, 2001).

29. Johannes R. Lischka, "Armor Plate: Nickel and Steel, Monopoly and Profit," *War, Business, and American Society: Historical Perspectives on the*

Military-Industrial Complex, Benjamin Franklin Cooling, ed. (Port Washington, NY: Kennikat Press, 1977), pp. 43–58.

30. David Kite Allison, *New Eye for the Navy: The Origin of Radar at the Naval Research Laboratory*, NRL Report 8466 (Washington, DC: Naval Research Laboratory, September 29, 1981).

31. *Federal Funds for Science XI: Fiscal Years 1961, 1962, and 1963*, NSF 63–11 (Washington, DC: National Science Foundation, 1963), Table C-32, p. 136. Government and industry together spent an estimated $345 million on R&D in 1940. Nearly 80 percent came from private firms; the federal government's contribution, military and nonmilitary, totaled only $74 million. John W. Kendrick, *Productivity Trends in the United States* (Princeton, NJ: Princeton University Press, 1961), p. 109.

32. Bush was a seminal figure; a considerable literature explores his career and influence. Proud to be an engineer and equally proud of his abilities as a bureaucratic operator, "Bush [established] a policy during the war of calling all OSRD members 'scientists,' in order to counter what he saw as the American military's antipathy toward engineering salesmanship and British snobbery toward engineering." Ronald Kline, "Construing 'Technology' as 'Applied Science': Public Rhetoric of Scientists and Engineers in the United States, 1880–1945," *Isis*, Vol. 86, 1995, pp. 194–221; quotation from p. 219.

33. Seven of the top ten contract recipients over OSRD's existence, 1941–1947, were universities. MIT headed the list with $117 million in awards; AT&T's Bell Laboratories, the leading corporate contractor, came fifth with a total of $16.4 million. Larry Owens, "The Counterproductive Management of Science in the Second World War: Vannevar Bush and the Office of Scientific Research and Development," *Business History Review*, Vol. 68, 1994, pp. 515–576: Appendix, p. 565.

34. "The major differences revolved around one broad question: Was the Navy to have its own 'army' and 'air force' and to decide for itself how large they were to be and how they were to be used?" Condit, *The History of the Joint Chiefs of Staff, Volume II*, p. 176. In the mid-1980s too, the Navy opposed reorganization proposals culminating in the Goldwater-Nichols Act more strenuously than the other services. Senator Goldwater himself reportedly telephoned the Pentagon's reorganization "crisis center," finding it staffed by Navy and Marine Corps officers at the ready to explain why his bill was a bad idea and unneeded. Michael Ganley, "How's That Again? You're Opposed to *What?*" *Armed Forces Journal International*, March 1986, p. 18.

35. Quoted in Michael S. Sherry, *Preparing for the Next War: American Plans for Postwar Defense, 1941–45* (New Haven, CT: Yale University Press, 1977), p. 150.

36. At its peak in 1958, the Soviet submarine fleet numbered about 475, including World War II hulls still in service. Norman Polmar and Kenneth J. Moore, *Cold War Submarines: The Design and Construction of*

U.S. and Soviet Submarines (Washington, DC: Brassey's, 2004), Table 1, p. xii, and p. 327.

37. Gerald Haines, "The National Reconnaissance Office: Its Origins, Creation, and Early Years," *Eye in the Sky: The Story of the CORONA Spy Satellites*, Dwayne A. Day, John M. Logsdon, and Brian Latell, eds. (Washington, DC: Smithsonian Institution Press, 1998), pp. 143–156; quotation from p. 144.

38. Ralph Sanders, "Introduction," *Technology, Strategy and National Security*, Franklin D. Margiotta and Ralph Sanders, eds. (Washington, DC: National Defense University Press, 1985), p. 4.

39. John T. Greenwood, "The Emergence of the Postwar Strategic Air Force, 1945–1953," *Air Power and Warfare: The Proceedings of the 8th Military History Symposium, United States Air Force Academy, 18–20 October 1978*, Alfred F. Hurley and Robert C. Ehrhart, eds. (Washington, DC: U.S. Government Printing Office, 1979), pp. 215–244; quotation from p. 229.

40. H.W. Brands, "The Age of Vulnerability: Eisenhower and the National Insecurity State," *American Historical Review*, Vol. 94, 1989, pp. 963–989.

41. See, e.g., David Alan Rosenberg, "The Origins of Overkill: Nuclear Weapons and American Strategy, 1945–1960," *International Security*, Vol. 7, No. 4, Spring 1983, pp. 3–71.

42. Mark Lorell, *The U.S. Combat Aircraft Industry, 1909–2000: Structure, Competition, Innovation* (Santa Monica, CA: RAND, 2003), p. 80.

43. Jacob Neufeld, *Ballistic Missiles in the United States Air Force, 1945–1960* (Washington, DC: Office of Air Force History, 1990), p. 87, notes a 1953 briefing chart that counted 28 guided missile programs run by one or another of the services. The Navy, with 12, had the most.

44. Summarizing retrospective studies of weapons programs, Ronald N. Kostoff, "Semiquantitative Methods for Research Impact Assessment," *Technological Forecasting and Social Change*, Vol. 44, 1993, pp. 231–244, writes, p. 234:

> The real difference in performance between a weapons system and its predecessor was usually not the consequence of one, two, or three scientific advances or technological capabilities but was the synergistic effect of 100, 200, or 300 advances, each of which alone was relatively insignificant.

Chapter 4   Understanding Acquisition

1. New York: Scribner's, 1988, p. 424. Lehman was Secretary of the Navy in the Reagan administration.

2. Alison Thomson, "Defense-Related Employment and Spending, 1996–2006," *Monthly Labor Review*, July 1998, pp. 14–33; *National Defense Budget Estimates for FY 2008* (Washington, DC: Office of the Under Secretary of Defense [Comptroller], March 2007), Table 7–6.

3. *The Long-Term Implications of Current Defense Plans: Detailed Update for Fiscal Year 2007* (Washington, DC: Congressional Budget Office, April 2007), Figure 3–2 Updated, p. 10. DOD-wide agencies such as the Defense Advanced Research Projects Agency and the Missile Defense Agency account for the remaining 10–11 percent (and more than one-quarter of RDT&E). The Army share of procurement has risen in the past several years because of the need to replace equipment lost or damaged in Afghanistan and Iraq.

4. [Captain] John Byron, "A New Navy for a New World," *U.S. Naval Institute Proceedings*, Vol. 129, March 2003, pp. 86–88; quotation from p. 87.

5. Franklin C. Spinney, "Defense Facts of Life," Staff Paper, U.S. Department of Defense, December 5, 1980, p. 97.

6. David A. Fulgum, "Boeing Military Battles for Profits with Technology," *Aviation Week & Space Technology*, May 24, 1999, pp. 76–77; Bruce Nordwall, "Aging Avionics Systems Bedevil Military Services," *Aviation Week & Space Technology*, June 25, 2001, p. 74.

7. [Lieutenant Commander] Eric Johns, "Perfect Is the Enemy of Good Enough," *U.S. Naval Institute Proceedings*, Vol. 114, October 1988, pp. 37–48; quotation from p. 37.

8. Robert Wall, "F-22 Enters Critical Phase," *Aviation Week & Space Technology*, June 24, 2002, pp. 48–49. The planned procurement had earlier been cut to 339. The Air Force changed the designation of the F-22 in 2002, calling it the F/A-22. The rebranding was intended to fend off critics by signaling expansion of the mission beyond air superiority to include ground attack. In 2006, with the plane safely in production, the Air Force reverted to the original designation.

9. "Program Acquisition Costs by Weapon System," Department of Defense Budget for Fiscal Year 2006, February 2005, p. 46.

10. *Defense Acquisitions: Key Decisions to Be Made on Future Combat System*, GAO-07-376 (Washington, DC: U.S. Government Accountability Office, March 2007), pp. 40–41.

11. "Resource Implications of the Navy's Fiscal Year 2008 Shipbuilding Plan," Congressional Budget Office, Washington, DC, March 23, 2007, pp. 13–14. This program began in 1998 as the DD(X). The cost figures are in 2008 dollars.

12. *The Long-Term Implications of Current Defense Plans: Detailed Update for Fiscal Year 2007*, Figure 3–1 Updated, p. 9. The projection, in 2007 dollars, includes cost growth based on historical experience.

13. *National Defense Budget Estimates for FY 2008* (Washington, DC: Office of the Under Secretary of Defense [Comptroller], March 2007), Table 1–10, p. 15. DOD normally projects five years ahead; thus 2012 is the last year included.

14. In the design of hull forms, for instance, experiment had begun to overcome guesswork and opinion only with the construction of towing tanks for tests on models, the first of which went into operation in Britain in

1872. David K. Brown, "The Era of Uncertainty, 1863–1878," *Steam, Steel and Shellfire: The Steam Warship 1815–1905*, Robert Gardiner, ed. (London: Conway Maritime Press, 1992), pp. 75–94.

15. John D. Anderson, Jr., *A History of Aerodynamics and Its Impact on Flying Machines* (Cambridge: Cambridge University Press, 1997) traces the slowly narrowing gap over many decades between advances in aeronautical theory and its applications to aircraft design.

16. The Signal Corps officer in charge of the Army's aircraft purchases, a man then in his fifties, had studied electrical engineering early in his career. Charles J. Gross, "George Owen Squier and the Origins of American Military Aviation," *Journal of Military History*, Vol. 54, 1990, pp. 281–305. When Jerome Hunsaker took over the Navy's Aircraft Division in 1916 as a young lieutenant (but not a pilot) and the only nonclerical staff member, he had just received MIT's first Ph.D. in aeronautics for wind tunnel studies of aircraft stability. William F. Trimble, *Jerome C. Hunsaker and the Rise of American Aeronautics* (Washington, DC: Smithsonian Institution Press, 2002), p. 39.

17. See, in general, Irving Brinton Holley, Jr., *United States Army in World War II, Special Studies—Buying Aircraft: Matériel Procurement for the Army Air Forces* (Washington, DC: Office of the Chief of Military History, Department of the Army, 1964), especially pp. 80–149. Also Jacob A. Vander Muelen, *The Politics of Aircraft: Building an American Military Industry* (Lawrence: University Press of Kansas, 1991), pp. 8–40.

18. Holley, *Buying Aircraft*, p. 512. He is discussing a later period, but the points are general.

19. Vander Muelen writes, *The Politics of Aircraft*, pp. 38–39:

    [P]roduction of the De Haviland observation plane . . . increased from approximately 1,800 in September 1918 to the incredible figure of 4,000 in November. However, this cumbersome craft only underscored the flawed approach taken by the Americans to providing themselves with air power. The DH-4, already obsolete in summer 1917, was built by Fisher Body and Dayton-Wright Corporation, which had been organized to mass-produce the aircraft for the war. The great resources poured into it were clearly driven more by the desires to get something—anything—to the front and so salvage as much of the program and various reputations as possible. So strong were these pressures that Dayton-Wright shipped DH-4s to the front after randomly testing one in six.

20. Walter G. Vincenti, *What Engineers Know and How They Know It: Analytical Studies from Aeronautical History* (Baltimore, MD: Johns Hopkins University Press, 1990) explores in depth and detail the blends of theory and empiricism characteristic of aircraft design during the interwar period.

21. Edward L. Homze, "The Luftwaffe's Failure to Develop a Heavy Bomber before World War II," *Aerospace Historian*, Vol. 24, No. 1, March 1977, pp. 20–26.

22. R.J. Overy, *The Air War, 1939–1945* (New York: Stein and Day, 1981), p. 120. The German total includes warheads carried by V-1 and V-2 missiles. The figures in the next paragraph come from the same source, pp. 123, 150, 172–173, and 178.

23. The phrase is from Perry McCoy Smith, *The Air Force Plans for Peace, 1943–1945* (Baltimore, MD: Johns Hopkins Press, 1970), p. 23.

24. Daniel R. Beaver, "'Deuce and a Half': Selecting U.S. Army Trucks, 1920–1945," *Feeding Mars: Logistics in Western Warfare from the Middle Ages to the Present*, John A. Lynn, ed. (Boulder, CO: Westview, 1993), pp. 251–270.

25. David E. Johnson, *Fast Tanks and Heavy Bombers: Innovation in the U.S. Army, 1917–1945* (Ithaca, NY: Cornell University Press, 1998), pp. 78–80.

26. Robert S. Cameron, "Armor Combat Development 1917–1945," *ARMOR*, Vol. 106, September–October 1997, pp. 15–19.

27. Christopher R. Gabel, "World War II Armor Operations in Europe," *Camp Colt to Desert Storm: The History of U.S. Armored Forces*, George F. Hofmann and Donn A. Starry, eds. (Lexington: University Press of Kentucky, 1999), pp. 144–184; quotation from p. 155.

28. Quoted on pp. 42–43 of Timothy K. Nenninger, "Organizational Milestones in the Development of American Armor, 1920–40," *Camp Colt to Desert Storm*, pp. 37–66.

29. Shelford Bidwell and Dominick Graham, *Fire-Power: British Army Weapons and Theories of War 1904–1945* (London: George Allen & Unwin, 1982) write, pp. 177–178:

> The three leading advocates of the armoured cause—Fuller, Liddell Hart and Hobart— . . . all shared one disabling habit of thought. Their reasoning was never empirical. They were all highly articulate and their polemics are full of vivid metaphors and similes. They were in short, *literary* in their approach . . . .

Emphasis in original.

30. As late as the mid-1930s, Chief of Staff Douglas MacArthur advocated a "machine-gun army." See pp. 123–124 in George F. Hofmann, "Army Doctrine and the Christie Tank: Failing to Exploit the Operational Level of War," *Camp Colt to Desert Storm*, pp. 92–143.

31. Thomas G. Mahnken, "Beyond Blitzkrieg: Allied Responses to Combined-Arms Armored Warfare during World War II," *The Diffusion of Military Technology and Ideas*, Emily O. Goldman and Leslie C. Eliason, eds. (Stanford, CA: Stanford University Press, 2003), pp. 243–266.

32. Most of the specifics in this section not otherwise cited come from Holley, *Buying Aircraft*.

33. John Perry Miller, *Pricing of Military Procurements* (New Haven, CT: Yale University Press, 1949), pp. 126–133.

34. Stuart D. Brandes, *Warhogs: A History of War Profits in America* (Lexington: University Press of Kentucky, 1997), Table 2, pp. 266–267, reports that the average pay of the presidents of 24 Wisconsin defense contractors

doubled between 1940 and 1945, then declined in 1946. Brandes also notes (photo caption between pp. 183 and 183) that Edwin and George Link, brothers who invented the famed Link trainer, a simulator in which pilots practiced, paid themselves salaries of $12,000 and $15,000 in 1939, rising to $72,000 and $90,000 in 1941 even as their patents brought them over $1 million in royalties.

35. The Army and Navy occasionally gave suppliers "experimental contracts," usually quite small, in effect for what would later be called R&D, when firms were unwilling to pay out of pocket for work the service wanted. Because this usually meant circumventing rules on competitive bidding, "the Army and Navy operated [in] continual fear of Congressional investigation of experimental contracts." Robert Schlaifer and S.D. Heron, *Development of Aircraft Engines [and] Development of Aviation Fuels: Two Studies of Relations between Government and Business* (Boston: Harvard University Graduate School of Business Administration, 1950), p. 555. As the authors explain elsewhere,

> [T]he Services have to gamble with public funds by opening experimental contracts to competitive bidding. In many cases of this kind the development engineer (but rarely the contracting officer) realizes that it is almost certain that the lower bidder will enter the lowest bid because he does not understand the problem and also because he would like some apparently profitable work in his shop (p. 624).
>
> The major and the only really important incentive leading private industry to carry out the development of aircraft engines effectively was the prospect of profits on quantity production.
>
> The services . . . permit[ted] the firms to count development of new models as current overhead and to include it in the cost and price of current models sold to the government (p. 9).

36. David A. Mindell, *Between Human and Machine: Feedback, Control, and Computing before Cybernetics* (Baltimore, MD: Johns Hopkins University Press, 2002), p. 42. That firms such as Ford Instrument Company and Carl Norden, Inc., the bombsight manufacturer, were "virtually a private extension" (p. 43) of the Navy, which might be their only customer, was evidently no obstacle and might indeed have made it easier for BuOrd to do as it wished.

37. Holley, *Buying Aircraft*, p. 147.

38. The predecessors of the $2000 coffee maker included the Navy's 11,000 dozen oyster forks, 95 percent "of such poor quality that . . . they were usable only in an emergency!" Cited in Charles J. Hitch and Roland N. McKean, *The Economics of Defense in the Nuclear Age* (Cambridge, MA: Harvard University Press, 1960), p. 51.

39. John E. Wiltz, *In Search of Peace: The Senate Munitions Inquiry, 1934–36* (Baton Rouge: Louisiana State University Press, 1963).

40. For a summary of 15 earlier studies, see *Assessing the Potential for Civil-Military Integration: Technologies, Processes, and Practices* (Washington, DC: Office of Technology Assessment, September 1994), pp. 171–175.

41. See the analysis of congressional voting patters in Kenneth R. Mayer, *The Political Economy of Defense Contracting* (New Haven, CT: Yale University Press, 1991).

42. That is the title of the first chapter in *Defense Reform Initiative Report* (Washington, DC: Office of the Secretary of Defense, November 1997).

43. *Directions for Defense: Report of the Commission on Roles and Missions of the Armed Forces* (Washington, DC: U.S. Government Printing Office, 1995), pp. 3–23, gives an estimate of 18 percent.

44. See, e.g., Mark Lorell, Jeffrey A. Drezner, and Julia Lowell, *Reforming Mil-Specs: The Navy Experience with Military Specifications and Standards Reform* (Santa Monica, CA: RAND, 2001). In one instance, a vendor charge of $15,000 for DOD-mandated screening of a batch of electronic circuit assemblies resulted in a price increase by more than 35 times (pp. 6–7). Required inspection practices may also do positive harm by introducing damage where none existed previously; 100 percent inspection of electronic components sometimes reduces reliability below that achieved through normal commercial practices of statistical process control.

45. Jacques Gansler, *Affording Defense* (Cambridge, MA: MIT Press, 1989), p. 191.

46. Steven Kelman, *Procurement and Public Management: The Fear of Discretion and the Quality of Government Performance* (Washington, DC: AEI Press, 1990), p. 34.

47. *Report of the Defense Science Board Task Force on Defense Software* (Washington, DC: Office of the Under Secretary of Defense for Acquisition and Technology, November 2000), p. 19.

Chapter 5   Organizing for Defense

1. Malcolm Moos wrote this widely reprinted speech, the source of several other well-known quotations.

2. Paul A.C. Koistinen, *Arsenal of World War II: The Political Economy of American Warfare, 1940–1945* (Lawrence, KS: University Press of Kansas, 2004), p. 54.

3. John Kenly Smith, Jr., "World War II and the Transformation of the American Chemical Industry," *Science, Technology, and the Military*, Vol. 2, Everett Mendelsohn, Merritt Roe Smith, and Peter Weingart, eds. (Dordrecht, Netherlands: Kluwer, 1988), pp. 307–322.

4. Gerald T. White, "Financing Industrial Expansion for War: The Origin of the Defense Plant Corporation Leases," *Journal of Economic History*, Vol. 9, No. 2, November 1949, pp. 156–183.

5. Margaret B.W. Graham and Bettye H. Pruitt, *R&D for Industry: A Century of Technical Innovation at Alcoa* (Cambridge: Cambridge University Press, 1990), pp. 468–469 and 491–492.

6. Bruce Brunton, "An Historical Perspective on the Future of the Military-Industrial Complex," *Social Science Journal*, Vol. 28, 1991, pp. 45–62.

7. On the politics underlying the shift from a mixed public-private defense base to near-total reliance on private firms, see Aaron L. Freidberg, *In the Shadow of the Garrison State: America's Anti-Statism and Its Cold War Grand Strategy* (Princeton, NJ: Princeton University Press, 2000), pp. 245–295. Freidberg stresses the ideological force of free-enterprise rhetoric and the contrasts drawn by those seeking to steer defense contracts to private firms with state control in Soviet Russia.

8. In the Pentagon's RDT&E nomenclature, category 6.1 is basic research, 6.2 applied research, and category 6.3 is designated advanced technology development (e.g., work aimed at establishing "proof-of-principle"). Together, these three categories comprise what DOD calls its Science & Technology (S&T) program. In recent years, the S&T program has accounted for about 20 percent of RDT&E spending. The other 80 percent, categories 6.4–6.7, covers the engineering of systems intended for production. Category 6.4 is labeled "advanced component development and prototypes" and 6.5 "system development and demonstration." Categories 6.6 and 6.7, "management support" and "operational system development," include overhead costs for test facilities, firing ranges, and proving grounds as well as expenditures for particular acquisition programs.

9. Beginning with the World War II Manhattan project, publicly available documents have included classified programs in reported budget totals while withholding information on the number of such programs and their funding. This means that the difference between total reported spending and the sum of spending on named programs should equal the amount spent on classified programs. According to Steven M. Kosiak, "Classified Funding in the FY 2007 Defense Budget Request," Center for Strategic and Budgetary Assessments, Washington, DC, May 17, 2006, classified programs accounted for about 17 percent of the fiscal 2006 acquisition budget. Because the Air Force is responsible for most military space activities, such as intelligence satellites, and many terrestrial communications systems likewise classified, over 40 percent of Air Force procurement in recent years has been "black." The Army and Navy conduct some secret RDT&E, but have had no classified procurements since the early 1990s.

10. While firms sometimes reap high profits on defense orders, and the perception has been widespread that this is common, there has never been much evidence of abnormally large profit margins as a general phenomenon. See, e.g., George J. Stigler and Claire Friedland, "Profits of Defense Contractors," *American Economic Review*, Vol. 61, 1971, pp. 692–694,

which found profitability in defense to be above that in other sectors of the economy during the 1950s, falling to levels around the industry-wide average in the following decade.

11. The list of Cold War bail-outs included Lockheed, Grumman, Litton Industries, and Chrysler (at the time the Army's favored source of tanks). Richard A. Stubbing with Richard A. Mendel, *The Defense Game* (New York: Harper & Row, 1986), pp. 36–39, 170, 184–190, and 214–215.

12. Scale economies result from efficiency-enhancing capital investments, such as automated production equipment. Learning economies stem primarily from "invisible" efficiency improvements as production workers find better ways to do their jobs. Early studies by economists of learning-in-production drew on World War II cost accounting records. From 1943 to 1946, for instance, Bath Iron Works built nearly 50 destroyers of essentially identical design. The first of these consumed some 1.5 million direct labor hours. At the end of the run, labor inputs had fallen by half. Because workforce turnover was low (meaning there were few newly hired employees to drag down productivity), because capital equipment did not change significantly, and because design modifications to the destroyers were inconsequential, the decline in labor inputs represents an almost pure example of learning economies. Henry A. Gemery and Jan S. Hogendorn, "The Microeconomic Bases of Short-Run Learning Curves: Destroyer Production in World War II," *The Sinews of War: Essays on the Economic History of World War II*, Geofrey T. Mills and Hugh Rockoff, eds. (Ames: Iowa State University Press, 1993), pp. 150–165.

13. Electric Boat, the low bidder, was originally to build the first batch of four Virginia-class hulls. The 1996 defense authorization act required the Navy to contract with Newport News Shipbuilding for two of these. Congress was doubly disingenuous in claiming that splitting the work would both preserve the "nuclear submarine industrial base" and save money through competition: the Navy predicted costs would rise under the imposed arrangement, as in fact they did. See *New Attack Submarine: Program Status*, GAO/NSIAD-97–25 (Washington, DC: U.S. General Accounting Office, December 1996), which warned not only of increases in overhead costs and reductions in learning economies, but that "The cost of transferring the submarine's design from the first to the second shipbuilder is based on an optimistic estimate that is less than the actual cost of the last major design transfer" (p. 2).

14. Harvey M. Sapolsky, "The Origins of the Office of Naval Research," *Naval History: The Sixth Symposium of the U.S. Naval Academy*, Daniel M. Masterson, ed. (Wilmington, DE: Scholarly Resources, 1987), pp. 206–225. The Army managed the Manhattan Project (Vannevar Bush ruled out the Navy as too difficult to work with, based on his experience earlier in the war). Afterward, Congress, believing the atomic bomb too

terrible to leave in the hands of the military, created the nominally civilian AEC to oversee both military and nonmilitary uses of nuclear energy. Successive reorganizations transmogrified the original five-person commission into the Department of Energy, which had defense programs budgeted at $9.6 billion in fiscal 2007 (with another $6 billion for cleanup of environmental hazards created in the past at weapons production and testing sites).

15. Institutional patterns diffused in considerable measure through personal networks, which at the time were relatively circumscribed since there were only a few research-intensive universities before the war and disciplinary fields in science and engineering were broader, with less of the specialization that followed in the 1960s and more permeable boundaries. Bush's friend and deputy Warren Weaver, for instance, had come to OSRD from the Rockefeller Foundation, where he had funded Bush's prewar work at MIT on differential analyzers (a form of analog computer), and after the war went on to serve as a member of ONR's civilian advisory board. G. Pascal Zachary, *Endless Frontier: Vannevar Bush, Engineer of the American Century* (New York: Free Press, 1997), pp. 73–74, 315.

16. Harvey M. Sapolsky, *Science and the Navy: The History of the Office of Naval Research* (Princeton, NJ: Princeton University Press, 1990), p. 44.

17. Congress attached the Mansfield amendment, directing DOD to restrict its R&D contracts to projects directly linked with military needs, to the 1970 and 1971 defense authorization bills. Easily evaded, the provision nonetheless made research managers in defense agencies cautious about supporting work in universities, where opposition to the Vietnam War had become intense and DOD-funded research drew frequent protests.

18. Sapolsky entitled one of his chapters in *Science and the Navy* "The Office of No Return?" after a jibe by ONR opponents. He suggests the Navy would have cast its new research arm overboard within a few years if the Korean War had not boosted the funds available for R&D (p. 120).

19. Quoted in Michael H. Gorn, *Harnessing the Genie: Science and Technology Forecasting for the Air Force, 1944–1986* (Washington, DC: Office of Air Force History, 1988), pp. 17–18.

20. Robert S. Leonard, Jeffrey A. Drezner, and Geoffrey Sommer, *The Arsenal Ship Acquisition Process Experience* (Santa Monica, CA: RAND, 1999).

21. Gorn, *Harnessing the Genie*, pp. 2, 59–60. Today, the Army, Navy, and Air Force each maintain scientific advisory bodies and the Defense Science Board serves DOD as a whole. Beginning in 1948, defense agencies also began sponsoring "summer studies" that brought together multidisciplinary teams of scientists and engineers, most of them academics and sometimes including social scientists, for problem-focused work on topics of interest to the military, along with summer schools that introduced younger scientists to defense issues and Pentagon patronage. The Navy organized the first summer study. J.R. Marvin and F.J. Weyl, "The

Summer Study," *Naval Research Reviews*, Vol. 19, No. 8, August 1966, pp. 1–7, 24–28.

22. Richard G. Hewlett and Francis Duncan, *Nuclear Navy, 1946–1962* (Chicago: University of Chicago Press, 1974).

23. Benjamin S. Kelsey, *The Dragon's Teeth? The Creation of United States Air Power for World War II* (Washington, DC: Smithsonian Institution Press, 1982), p. 43.

24. Robert Schlaifer and S.D. Heron, *Development of Aircraft Engines [and] Development of Aviation Fuels: Two Studies of Relations between Government and Business* (Boston: Harvard University Graduate School of Business Administration, 1950), pp. 11 and 35.

25. Quoted in Edward H. Sims, *Fighter Tactics and Strategy, 1914–1970* (Fallbrook, CA: Aero Publishers, 1980), p. 244.

26. Stephen B. Johnson, *The United States Air Force and the Culture of Innovation: 1945–1965* (Washington, DC: Air Force History and Museums Program, 2002), p. 11.

27. Jacob Neufeld, *Ballistic Missiles in the United States Air Force, 1945–1960* (Washington, DC: Office of Air Force History, 1990), pp. 111–128; Thomas P. Hughes, *Rescuing Prometheus* (New York: Pantheon, 1998), pp. 69–139.

28. The Navy's use or nonuse of PERT—the acronym stands for Program Evaluation [and] Review Technique—in the Polaris program has become one of the well-known episodes in the spread of systems management. Harvey M. Sapolsky, in *The Polaris System Development: Bureaucratic and Programmatic Success in Government* (Cambridge, MA: Harvard University Press, 1972), describes PERT as "An alchemous combination of whirling computers, brightly colored charts, and fast-talking public relations officers . . . . [T]he [Polaris] Office won its battle for funds and priority. Its program was protected from the bureaucratic interference of the comptrollers and the auditors" (p. 129). "It mattered not how PERT was used, only that it was in use" (p. 124). Whether or not it contributed much to Polaris beyond insulation against micro-managing politicians and bureaucrats, by the early 1960s PERT had become a commonplace tool of project and program management, in the civilian economy as well as DOD.

29. The Navy's Sidewinder missile, developed in the early 1950s, is perhaps the best-known example of a bootleg military project. See pp. 315–320 in David K. Allison, "U.S. Navy Research and Development since World War II," *Military Enterprise and Technological Change: Perspectives on the American Experience*, Merritt Roe Smith, ed. (Cambridge, MA: MIT Press, 1985), pp. 289–328.

30. For other examples of innovations fostered by networks of officers in the intermediate ranks who begin by securing a few well-placed supporters able to help them get the funds needed to demonstrate feasibility and then go on to win the approvals necessary for full-scale development, see Vincent

Davis, *The Politics of Innovation: Patterns in Navy Cases*, Monograph Series in World Affairs, Vol. 4, Monograph No. 3 (Denver, CO: University of Denver, Social Science Foundation and Graduate School of International Studies, 1967). Much the same thing happens in private firms.

31. ARPA became the Defense Advanced Research Projects Agency, DARPA, in 1972, a change intended to signal its focus on military research at a time of congressional distress, as expressed by the Mansfield amendment, over DOD funding that strayed, or might be seen to stray, from clear-cut military needs. In 1993, the Clinton administration, viewing the agency as a vehicle for dual-use R&D, restored the original name. Three years later a Republican Congress piqued by ARPA's ventures into commercially oriented technologies put "Defense" back in. Hereafter it will be referred to simply as DARPA.

32. T. Keith Glennan, as cited in Robert Frank Futrell, *Ideas, Concepts, Doctrine: Basic Thinking in the United States Air Force, 1907–1960,* Vol. I (Maxwell Air Force Base, AL: Air University Press, December 1989), p. 594.

33. Virginia P. Dawson, *Engines and Innovation: Lewis Laboratory and American Propulsion Technology*, NASA SP-4306 (Washington, DC: National Aeronautics and Space Administration, 1991).

34. "Swept-back wings, delta wings, wings with variable sweep-back, leading-edge flaps—all came from Germany during the war." Ronald Miller and David Sawers, *The Technical Development of Modern Aviation* (London: Routledge & Kegan Paul, 1968), p. 173.

35. In a fly-by-wire (FBW) control system, computers and electronics replace electro-mechanical components, permitting trim adjustments at much faster rates and with far greater discrimination than the best human pilots. See, e.g., James E. Tomayko, *Computers Take Flight: A History of NASA's Pioneering Digital Fly-By-Wire Project*, NASA SP-2000-4224 (Washington, DC: National Aeronautics and Space Administration, 2000). After initial development for the Apollo program, FBW was applied to the F-16 and then the F-117. With its faceted surfaces to reduce radar reflections, the aerodynamically unstable F-117 was sometimes called the "hopeless diamond" during conceptual design; such a plane could not stay aloft without computer controls. FBW offers further advantages to the military, including reduced vulnerability to enemy fire. During the Vietnam War, the United States lost many planes to ground fire that damaged hydraulics and caused loss of control. Electronic components, much smaller, are less likely to be hit.

36. By the time the Clinton administration canceled NASP in 1993, projected spending had reached $17 billion, five times the figure put forward when the Reagan administration approved the program in 1987. See, e.g., Bob Davis and Andy Paztor, "Blue Sky: Plane of the Future Would Fly into Orbit, Cross U.S. in an Hour," *Wall Street Journal*, June 19, 1989, pp. A1, A7; and Stanley W. Kandebo, "NASP Canceled, Program Redirected," *Aviation Week & Space Technology*, June 14, 1993, pp. 33–34.

37. Futrell, *Ideas, Concepts, Doctrine: Basic Thinking in the United States Air Force, 1907–1960,* Vol. I, pp. 477–504.

38. On the composition of DARPA R&D over the years, see Richard H. Van Atta, Seymour J. Deitchman, and Sidney G. Reed, "DARPA Technical Accomplishments, Volume III: An Overall Perspective and Assessment of the Technical Accomplishments of the Defense Advanced Research Projects Agency: 1958–1990," IDA Paper P-2538, Institute for Defense Analyses, Alexandria, VA, July 1991. Although DARPA has supported a wide range of work, including stealth, three areas stand out: information technologies, directed energy weapons (e.g., lasers for ballistic missile defense), and sensors and surveillance. DARPA-sponsored work in the last of these categories, for instance, led to the electronic imaging systems installed in place of periscopes on Virginia-class submarines (because periscopes fit awkwardly within a submarine hull, this opens up design configurations earlier precluded).

39. R. Hatch, Joseph L. Luber, and James H. Walker, "Fifty Years of Strike Warfare Research at the Applied Physics Laboratory," *Johns Hopkins APL Technical Digest*, Vol. 11, No. 1, 1992, pp. 113–123.

40. An "innovation system" is an analytical construct intended to help weigh the relative contributions of institutions including education and training, intellectual property protection, capital, labor, and product markets, and R&D. Richard R. Nelson, ed., *National Innovation Systems: A Comparative Analysis* (New York: Oxford University Press, 1993) provides an introduction. While the literature on national systems of innovation continues to expand, military technology has received relatively little attention beyond obligatory references to levels of R&D spending and occasional discussions of spin-off. The reasons begin with the recondite nature of high-technology weaponry, so unlike the commercial technologies that most analysts of innovation explore.

Chapter 6   Buying Military Systems: What the Services Want and What They Get

1. *Reflections on Research and Development in the United States Air Force: An interview with General Bernard A. Schriever and Generals Samuel C. Phillips, Robert T. March, and James H. Doolittle, and Dr. Ivan A. Getting, Conducted by Dr. Richard H. Kohn,* Jacob Neufeld, ed. (Washington, DC: Center for Air Force History, 1993), p. 83. Getting, an alumnus of the World War II Office of Scientific Research and Development and the MIT Radiation Laboratory who became president of the Aerospace Corporation, is perhaps best known as one of the developers of the Global Positioning System.

2. John P. Crecine, "Defense Resource Allocation: Garbage Can Analysis of C^3 Procurement," *Ambiguity and Command: Organizational Perspectives on Military Decision Making*, James G. March and Roger Weissinger-Baylon, eds. (Marshfield, MA: Pitman, 1986), pp. 72–119; quotation from p. 102.

3. The labyrinthine structure of acquisition is described in *Introduction to Defense Acquisition Management*, Seventh Edition (Fort Belvoir, VA: Defense Acquisition University Press, September 2005).

4. *Best Practices: Better Support of Weapon System Program Managers Needed to Improve Outcomes*, GAO-06-110 (Washington, DC: U.S. Government Accountability Office, November 2005), p. 5.

5. Robert S. Gilmour and Eric Minkoff, "Producing a Reliable Weapons System: The Advanced Medium-Range Air-to-Air Missile (AMRAAM)," *Who Makes Public Policy? The Struggle for Control between Congress and the Executive*, Robert S. Gilmour and Alexis A. Halley, eds. (Chatham, NJ: Chatham House, 1994), pp. 195–218; quotation from p. 206.

6. *Quadrennial Defense Review Report* (Washington, DC: Department of Defense, February 6, 2006), pp. 70–71.

7. *Best Practices: Better Support of Weapon System Program Managers Needed to Improve Outcomes*, p. 51, quoting from one of the many interviews with senior DOD managers (and surveys) on which this report is based.

8. Rumsfeld quotations from p. 38 of *Beyond Goldwater-Nichols: Defense Reform for a New Strategic Era, Phase 1 Report* (Washington, DC: Center for Strategic and International Studies, March 2004). The quoted phrase in the following sentence is from the text of the report itself, also p. 38.

9. Wilson to Neil H. McElroy, as quoted in James M. Roherty, *Decisions of Robert S. McNamara: A Study of the Role of the Secretary of Defense* (Coral Gables, FL: University of Miami Press, 1970), p. 50.

10. Quoted in William W. Kaufmann, *The McNamara Strategy* (New York: Harper & Row, 1964), pp. 172–173.

11. David C. Jones, "Comments on Smith's Proposals," *Reorganizing America's Defense: Leadership in War and Peace*, Robert J. Art, Vincent Davis, and Samuel P. Huntington, eds. (McLean, VA: Pergamon-Brassey's, 1985), pp. 330–343; quotation from p. 340.

12. Quoted in Michael Barzelay and Colin Campbell, *Preparing for the Future: Strategic Planning in the U.S. Air Force* (Washington, DC: Brookings, 2003), p. 60.

13. Harvey M. Sapolsky, *The Polaris System Development: Bureaucratic and Programmatic Success in Government* (Cambridge, MA: Harvard University Press, 1972). Also see Graham Spinardi, *From Polaris to Trident: The Development of US Fleet Ballistic Missile Technology* (Cambridge: Cambridge University Press, 1994).

14. Paul G. Kaminski, "Low Observables: The Air Force and Stealth," *Technology and the Air Force: A Retrospective Assessment*, Jacob Neufeld, George M. Watson, Jr., and David Chenoweth, eds. (Washington, DC: Air Force History and Museums Program, 1997), pp. 299–309. While a number of earlier aircraft, going back at least to the U-2, included design features intended to reduce radar returns, and the SR-71 Blackbird was designed from the beginning for low radar cross-section, the F-117 was the first design intended for volume production and routine service.

15. Sapolsky, *The Polaris System Development*, pp. 53 and 18. The author also provides this glimpse of interservice rivalry (p. 39):

> In late 1957, the Air Force, stating that it might have a land-based requirement for the Polaris, requested the Navy to prepare a study of possible system characteristics . . . . The Navy . . . turned over extensive information on the advantages and disadvantages of Polaris . . . . The land-based application did not live much longer, but, since the Navy was already publicizing all of the advantages . . . , the Air Force thought it was necessary to reveal . . . the disadvantages . . . .

16. To reduce costs, "Compared with the Seawolf, the [Virginia] is slower, carries fewer weapons, and is less capable in diving depth and arctic operations." *New Attack Submarine: Program Status*, GAO/NSIAD-97–25 (Washington, DC: U.S. General Accounting Office, December 1996), p. 11.

17. Statement of J. Michael Gilmore, Assistant Director, and Eric J. Labs, Principal Analyst, Congressional Budget Office, "Potential Costs of the Navy's 2006 Shipbuilding Plan," before the Subcommittee on Projection Forces, Committee on Armed Services, U.S. House of Representative, March 30, 2006, p. 10. The estimate is in 2007 dollars and excludes RDT&E and related expenditures through 2005.

18. Wayne P. Hughes, Jr., *Fleet Tactics and Coastal Combat*, Second Edition (Annapolis, MD: Naval Institute Press, 2000), p. 167.

19. Norman Friedman, *U.S. Submarines since 1945: An Illustrated Design History* (Annapolis, MD: Naval Institute Press, 1994), pp. 209–211.

20. [Rear Admiral] Al Konetzni, "How Many Subs Do We Need?" *U.S. Naval Institute Proceedings*, Vol. 126, November 2000, pp. 56–57; quotation from p. 57.

21. Robert T. Finney, *History of the Air Corps Tactical School 1920–1940* (Washington, DC: Center for Air Force History, 1992 [1955]), p. 68.

22. Finney, *History of the Air Corps Tactical School 1920–1940*, pp. 73–74.

23. Chennault retired in 1937, then in World War II led the "Flying Tigers"—officially, the American Volunteer Group—against the Japanese for Chiang Kai-shek. The insularity of the interwar Air Corps stemmed in part from the small size of its officer corps. "From 1921 to 1938, if not beyond, the same 75 to 90 field-grade officers [major and above] . . . controlled the inner workings of the air arm." Richard G. Davis, *Carl A. Spaatz and the Air War in Europe* (Washington, DC: Center for Air Force History, 1993), p. 10. These men, pilots determined to carve out a place for air power within a mostly uncomprehending army, came to think largely alike.

24. Shelford Bidwell and Dominick Graham, *Fire-Power: British Army Weapons and Theories of War 1904–1945* (London: George Allen & Unwin, 1982), p. 214.

25. Quoted on p. 251 in W.A. Jacobs, "The Battle for France, 1944," *Case Studies in the Development of Close Air Support*, Benjamin Franklin Cooling, ed. (Washington, DC: Office of Air Force History, 1990), pp. 237–293.

26. The quoted phrase, from a directive for the signature of Carl Spaatz prepared by Hoyt Vandenberg, appears on p. 10 in I.B. Holley, Jr., "Of Saber Charges, Escort Fighters, and Spacecraft: The Search For Doctrine," *Air University Review*, Vol. 34, No. 6, September–October 1983, pp. 2–11. Spaatz would become the first chief of staff of the independent postwar Air Force, Vandenberg his successor.

27. James Parton, "The Thirty-One Year Gestation of the Independent USAF," *Aerospace Historian*, Vol. 34, 1987, pp. 151–157; quotation from p. 153.

28. Ian Gooderson, *Air Power at the Battlefront: Allied Close Air Support in Europe, 1943–1945* (London: Cass, 1998), p. 227. Thomas Alexander Hughes, in *Over Lord: General Pete Quesada and the Triumph of Tactical Air Power in World War II* (New York: Free Press, 1995), writes that the Army Air Forces began World War II lacking "a clear and articulated doctrine of tactical air power," had to "devise makeshift plans and procedures for air support," and then found "they also lacked an analytical framework within which to evaluate the success of their tactical efforts after the event" (p. 253). Hughes's own narrative makes "triumph" (in his subtitle) seem a considerable overstatement; what happened might be better summarized as a long, slow, painful, if ultimately successful, recovery from initial failures in preparation.

29. The U.S. firm North American Aviation originally designed the famed P-51 Mustang in 1940 at the request of the Royal Air Force. The U.S. Army Air Forces did not like the plane and only in "the latter half of 1942 . . . finally recognized [it as] the best available fighter airframe . . . ." Robert Schlaifer and S.D. Heron, *Development of Aircraft Engines [and] Development of Aviation Fuels: Two Studies of Relations between Government and Business* (Boston: Harvard University Graduate School of Business Administration, 1950), p. 310.

30. See, e.g., Wesley Frank Craven and James Lea Cate, eds., *The Army Air Forces in World War II: Volume Three, Europe: Argument to V-E Day, January 1944 to May 1945* (Washington, DC: Office of Air Force History, 1983 [reprint of 1951 University of Chicago Press edition]), the official history, which portrays a relentlessly single-minded commitment to strategic bombing by the high command.

31. Quoted on p. 84 of Philip S. Meilinger, "The Admirals' Revolt of 1949: Lessons for Today," *Parameters*, Vol. 19, September 1989, pp. 81–96.

32. [Colonel] Alan L. Gropman, "Discussion and Comments," *Military Planning in the Twentieth Century, Proceedings of the Eleventh Military History Symposium, 10–12 October 1984*, Harry R. Borowski, ed. (Washington, DC: Office of Air Force History, 1986), pp. 244–248; quotation from p. 244. The F-105 was "fast and stable, a machine that pilots called 'honest' . . . well-designed for the single purpose of fighting a nuclear war." Kenneth P. Werrell, "Did USAF Technology Fail in Vietnam? Three Case Studies," *Airpower Journal*, Vol. 12, No. 1, Spring 1998, pp. 87–99; quotation from p. 88. Pilots called the plane the "Ultra Hog," among other nicknames alluding to the flight characteristics that handicapped it in air-to-air combat.

33. Richard P. Hallion, "A Troubling Past: Air Force Fighter Acquisition since 1945," *Airpower Journal*, Vol. 4, No. 4, Winter 1990, pp. 4–23; quotation from p. 9.

34. Robert Frank Futrell, *Ideas, Concepts, Doctrine: Basic Thinking in the United States Air Force, 1961–1984,* Vol. II (Maxwell Air Force Base, AL: Air University Press, December 1989), pp. 287–298.

35. James P. Stevenson, *The Pentagon Paradox: The Development of the F-18 Hornet* (Annapolis, MD: Naval Institute Press, 1993), Table I-3, p. 10.

36. Clark G. Reynolds, *The Fast Carriers: The Forging of an Air Navy* (New York: McGraw-Hill, 1968), p. 344.

37. David A. Fulgum, "B-2 Buy Tangled in Mission Rivalries," *Aviation Week & Space Technology*, January 16, 1995, pp. 61–62.

38. *Defense Science Board Summer Study on Transformation: A Progress Assessment, Volume I* (Washington, DC: Office of the Under Secretary of Defense for Acquisition, Technology, and Logistics, February 2006), p. 15.

39. Futrell, *Ideas, Concepts, Doctrine: Basic Thinking in the United States Air Force, 1961–1984,* Vol. II, p. 298. Navy pilots ordinarily flew somewhat less hazardous missions.

40. See, e.g., Admiral Bill Owens with Ed Offley, *Lifting the Fog of War* (New York: Farrar, Straus and Giroux, 2000), pp. 158–159 and 164–165.

41. The definitive treatment is Thomas L. McNaugher, *The M16 Controversies: Military Organizations and Weapons Acquisition* (New York: Praeger, 1984). Also see Edward C. Ezell, "Patterns in Small-Arms Procurement since 1945: Organization for Development," *War, Business, and American Society: Historical Perspectives on the Military-Industrial Complex*, Benjamin Franklin Cooling, ed. (Port Washington, NY: Kennikat Press, 1977), pp. 146–157.

42. The account that follows is based primarily on Nick Kotz, *Wild Blue Yonder: Money, Politics, and the B-1 Bomber* (New York: Pantheon, 1988).

43. James G. Burton, *The Pentagon Wars: Reformers Challenge the Old Guard* (Annapolis, MD: Naval Institute Press, 1993), pp. 31–35, suggests that the Air Force, knowing the B-1 was not a good airplane, did not fight hard against Carter's cancellation of the program. Burton (a former Air Force officer) writes, "I think the generals were happy to have this albatross removed from their collective neck. They would welcome the B-1 back when President Reagan later showed up with his moneybags" (p. 35).

44. *The B-1B Bomber and Options for Enhancements* (Washington, DC: Congressional Budget Office, August 1988).

45. George Sammet, Jr. and David E. Green, *Defense Acquisition Management* (Boca Raton: Florida Atlantic University Press, 1990), pp. 3–4.

46. David C. Aronstein and Albert C. Piccirillo, "The F-16 Lightweight Fighter: A Case Study in Technology Transition," *Technology and the Air Force: A Retrospective Assessment*, Jacob Neufeld, George M. Watson, Jr., and David Chenoweth, eds. (Washington, DC: Air Force History and Museums Program, 1997), pp. 203–229; quotation from p. 207. The

Navy's F/A-18 stemmed from Northrop's losing entry in the Air Force lightweight fighter competition.

47. James Fallows, *National Defense* (New York: Random House, 1981), provides a broad treatment of the 1970s military reform movement. Also see Mary Kaldor, *The Baroque Arsenal* (New York: Hill and Wang, 1981). The reformers argued for simpler weapons costing less that could be purchased in greater numbers. Many also held that reduced complexity would yield superior combat performance. Although reform advocates managed a partial success with the F-16—the final design drew criticism as compromising on the original objectives—the larger set of objections to "baroque" technology had little staying power.

48. *The Long-Term Implications of Current Defense Plans* (Washington, DC: Congressional Budget Office, January 2003), p. 45.

Chapter 7   Research and Weapons Design

1. Quoted on p. 83 of David K. Brown, "The Era of Uncertainty, 1863–1878," *Steam, Steel and Shellfire: The Steam Warship 1815–1905*, Robert Gardiner, ed. (London: Conway Maritime Press, 1992), pp. 75–94.

2. George Sammet, Jr. and David E. Green, *Defense Acquisition Management* (Boca Raton: Florida Atlantic University Press, 1990), p. 299.

3. This section draws in places on John A. Alic, "Policies for Innovation: Learning from the Past," *The Role of Government in Technology Innovation: Insights for Government Policy in the Energy Sector*, Vicki Norberg-Bohm, ed., Energy Technology Innovation Project, John F. Kennedy School of Government, Harvard University, October 2002, pp. 21–35.

4. "In retrospect, the development of nylon appears to be the solution of thousands of small problems . . . ." David A. Hounshell and John Kenly Smith, Jr., *Science and Corporate Strategy: Du Pont R&D, 1902–1980* (Cambridge: Cambridge University Press, 1988), p. 261. As discovery is taken to mark invention, marketplace introduction—commercialization—is usually taken to mark innovation.

5. William Aspray, "The Intel 4004 Microprocessor: What Constituted Invention?" *IEEE Annals of the History of Computing*, Vol. 19, No. 3, 1997, pp. 4–15. Texas Instruments built a microprocessor about the same time, but did not commercialize it.

6. See, e.g., John A. Alic, Lewis M. Branscomb, Harvey Brooks, Ashton B. Carter, and Gerald L. Epstein, *Beyond Spinoff: Military and Commercial Technologies in a Changing World* (Boston: Harvard Business School Press, 1992), pp. 350–354, and the sources cited therein.

7. See, e.g., Robert E. Cole, "Market Pressures and Institutional Forces: The Early Years of the Quality Movement," *The Quality Movement & Organization Theory*, Robert E. Cole and W. Richard Scott, eds. (Thousand Oaks, CA: Sage, 2000), pp. 67–87.

8. Quoted in Paolo E. Coletta, *The United States Navy and Defense Unification, 1947–1953* (Newark: University of Delaware Press, 1981), p. 69.

9. The most authoritative and influential of these surveys was *Current Industrial Reports: Manufacturing Technology 1988* (Washington DC: U.S. Department of Commerce, Bureau of the Census, May 1989).

10. Off Okinawa in the spring of 1945, Japan sent 1,900 kamikaze pilots against more than 1,200 American ships, sinking 26 and seriously damaging many more. In some 4,000 conventional attacks, Japanese pilots sank only 10 U.S. ships. George W. Baer, *One Hundred Years of Sea Power: The U.S. Navy, 1890–1990* (Stanford, CA: Stanford University Press, 1994), p. 266. Japan resorted to kamikaze tactics as it grew short of capable planes and pilots to fly them; some pilots had less than 10 hours of flight time when they took off on their first and final mission.

11. The account that follows is based on Malcolm Muir, Jr., *Black Shoes and Blue Water: Surface Warfare in the United States Navy, 1945–1975* (Washington, DC: Naval Historical Center, 1996), pp. 53ff.

12. Quotation from the flotilla commander's report in Muir, *Black Shoes and Blue Water*, p. 138.

13. "U.S. Navy Fact File: Aegis Weapons System," Naval Sea Systems Command (OOD), Washington, DC, March 29, 2006. Some sources give tracking capacities of many hundreds of targets. Aegis illustrates the automated implementation of tactical doctrine: computer software assigns priorities to each threat detected by the system's radar; while those priorities can be manually overridden, this takes time that might not be available in combat and risks distracting crew members from other vital tasks.

14. Admiral Thomas R. Weschler recalled, "I was on the source selection board to pick the Aegis contractor in the spring of 1967. Then we saw it go to sea in [the missile cruiser] Ticonderoga in 1985, about 18 years later." Quoted in Muir, *Black Shoes and Blue Water*, p. 231.

15. John A. Alic, "Science, Technology, and Economic Competitiveness," *Commercializing High Technology: East and West*, Judith B. Sedaitis, ed. (Lanham, MD: Rowman & Littlefield, 1997), pp. 3–18.

16. One of the principals described this research as "simply cut-and-try, because there was no theoretical guidance to work from." William F. Bahret, "The Beginnings of Stealth Technology," *IEEE Transactions on Aerospace and Electronic Systems*, Vol. 29, 1993, pp. 1377–1385; quotation from p. 1380.

17. *Report of the Defense Science Board Task Force on Acquisition Reform: Phase IV, Subpanel on Research and Development* (Washington, DC: Office of the Under Secretary of Defense for Acquisition and Technology, July 1999), Figure II-B, p. 4.

18. Elting E. Morison, *Men, Machines, and Modern Times* (Cambridge, MA: MIT Press, 1966), p. 105.

19. This section draws on John A. Alic, "Computer Assisted Everything? Tools and Techniques for Design and Production," *Technological Forecasting and Social Change*, Vol. 44, 1993, pp. 359–374.

20. The point is routinely noted in monographs and reviews summarizing the state-of-the-art in particular fields. For rocket motors, "A [mathematical] model to select the optimum system for any application would be impossible." Ben F. Wilson, "Liquid Rockets," *Tactical Missile Propulsion*, Gordon E. Jensen and David W. Netzer, eds. (Reston, VA: American Institute of Aeronautics and Astronautics, 1996), pp. 33–55; quotation from p. 37. The author does not bother to explain this statement; he simply issues the standard reminder.

21. Leon Trilling, "Styles of Military Technology Development: Soviet and U.S. Jet Fighters, 1945–1960," *Science, Technology, and the Military*, Vol. 1, Everett Mendelsohn, Merritt Roe Smith, and Peter Weingart, eds. (Dordrecht, Netherlands: Kluwer, 1988), pp. 155–186. Heavy losses in World War II were one reason the Soviet Union built so many fighters during the Cold War—a cumulative total of more than 50,000, twice as many as produced by the United States.

22. *Best Practices: A More Constructive Test Approach Is Key to Better Weapon System Outcomes*, GAO/NSIAD-00-199 (Washington, DC: U.S. General Accounting Office, July 2000), pp. 5 and 6.

23. For a brief account of DIVAD, see David C. Morrison, "The Beast Untamed," *National Journal*, November 18, 1989, pp. 2820–2822.

24. "Prepared Statement of Anthony R. Battista, professional staff member, House Committee on Armed Services," *Department of Defense Test Procedures*, hearing, Research and Development Subcommittee, Committee on Armed Services, House of Representatives, January 28, 1986 (Washington, DC: U.S. Government Printing Office, 1986), pp. 27–39; quotation from p. 28.

25. On the long-running battle first to persuade the Army to conduct realistic tests of the Bradley's armor and, once testing had revealed that the armor provided little protection, to modify it, see James G. Burton, *The Pentagon Wars: Reformers Challenge the Old Guard* (Annapolis, MD: Naval Institute Press, 1993), pp. 126–212.

26. This was the ratio around 1980. The sizes of armored forces varied considerably over the four-plus decades of the Cold War and the "balance" also depended on whether the tanks of other members of the North Atlantic Treaty Organization and Warsaw Pact were included in the comparison.

27. Donald MacKenzie, *Inventing Accuracy: A Historical Sociology of Nuclear Missile Guidance* (Cambridge, MA: MIT Press, 1990).

Chapter 8   Complex Systems and the Revolution in Military Affairs

1. Translated by Louise and Aylmer Maude (New York: Oxford University Press, 1998), p. 760.

2. Benjamin S. Lambeth, *The Transformation of American Air Power* (Ithaca, NY: Cornell University Press, 2000), p. 152. The aircraft types included

long-range bombers and short-range attack planes, tankers, AWACS (airborne warning and control system) and JSTARS (joint surveillance target attack radar system) planes, and the air superiority fighters that protected the rest. (AWACS planes "manage" air operations. JSTARS planes carry specialized equipment for picking out targets on the ground; their highly sensitive radars can, for instance, detect the characteristic signatures of tracked and treaded vehicles, enabling tanks to be distinguished from trucks.)

3. Quoted on p. 24 of Eric J. Lerner, "Lessons of Flight 655," *Aerospace America*, April 1989, pp. 18–26.

4. Gene I. Rochlin, "Iran Air Flight 655 and the USS Vincennes: Complex, Large-Scale Military Systems and the Failure of Control," *Social Responses to Large Technical Systems: Control or Anticipation*, Todd R. La Porte, ed. (Dordrecht, Netherlands: Kluwer, 1991), pp. 99–125.

5. [Lieutenant Commander] Eric Johns, "Perfect Is the Enemy of Good Enough," *U.S. Naval Institute Proceedings*, Vol. 114, No. 10, October 1988, pp. 37–48; quotation from p. 38.

6. Ronald H. Spector, *At War at Sea: Sailors and Naval Combat in the Twentieth Century* (New York: Viking, 2001), pp. 388–389.

7. *Effectiveness Of U.S. Forces Can Be Increased through Improved Weapon System Design*, PSAD-81-17 (Washington, DC: U.S. General Accounting Office, January 29, 1981), p. 27.

8. Human factors engineering, which deals with the behavior of operators as part of some larger system, arose in part from studies of "pilot error" during World War II and afterward. Investigators found case after case of confusing and counterintuitive instruments and control layouts, often traceable to design choices made for electro-mechanical convenience. For example, the B-25C "trim tab control trims in the opposite direction for which the correction is desired. Because [of this] opposite trim was put in, giving the aircraft a tendency to snap roll. An accident was narrowly averted." Pilot quotation from Paul M. Fitts and R.E. Jones, "Analysis of Factors Contributing to 460 'Pilot-Error' Experiences in Operating Aircraft Controls," Memorandum Report TSEAA-694-12, Air Force Aero Medical Laboratory, July 1, 1947, p. 25, reprinted in *Selected Papers on Human Factors in the Design and Use of Control Systems*, H. Wallace Sinaiko, ed. (New York: Dover, 1961), pp. 332–358.

9. The estimate, for fiscal 2003, is reported in *Defense Acquisitions: Stronger Management Practices Are Needed to Improve DOD's Software-Intensive Weapon Acquisitions*, GAO-04–393 (Washington, DC: U.S. General Accounting Office, March 2004), p. 1.

10. The discussion that follows draws in part on John A. Alic, Jameson R. Miller, and Jeffrey A. Hart, "Computer Software: Strategic Industry," *Technology Analysis & Strategic Management*, Vol. 3, 1991, pp. 177–190; and on John A. Alic, "Computer Assisted Everything? Tools and Techniques for Design and Production," *Technological Forecasting and Social Change*,

Vol. 44, 1993, pp. 359–374. Several otherwise uncited examples are based on presentations or discussion at the symposium on "Software, Growth, and the Future of the U.S. Economy," Board on Science, Technology, and Economic Policy, National Academy of Sciences, Washington, DC, February 20, 2004.

11. It took more than two centuries for the real price of lighting to fall by a similar amount, according to estimates in William D. Nordhaus, "Do Real-Output and Real-Wage Measures Capture Reality? The History of Lighting Suggests Not," *The Economics of New Goods*, Timothy J. Bresnahan and Robert J. Gordon, eds. (Chicago: University of Chicago Press, 1997), pp. 29–66. Multiple "revolutions" in lighting, with new sources of illumination replacing old—gas instead of candles and whale oil, followed by incandescent and then fluorescent lamps—were necessary for order-of-magnitude gains. In some contrast, advances in computing have resulted almost exclusively from a continuous stream of incremental improvements. Perhaps the only innovation in computing as fundamental as the shift from, say, gas light to electricity was the stored program, an invention that dates from the earliest days of digital processing and one that might more appropriately be taken as the starting point of the computer "revolution" than as a step change punctuating a longer sequence of technological advance.

12. Over several decades, real prices for software have declined at perhaps 2–3 percent annually, compared with 15–20 percent for hardware. See, e.g., pp. 189–192 in Martin Campbell-Kelly, "Software as an Economic Activity," *History of Computing: Software Issues*, Ulf Hashagen, Reinhard Keil-Slawik, and Arthur Norberg, eds. (Berlin: Springer, 2002), pp. 185–202, and the commentaries that follow.

13. ENIAC, the first electronic digital computer, was built for the U.S. Army's Ballistic Research Laboratory to automate compilation of standardized artillery firing tables. These relate the range of a projectile to variables such as barrel elevation, bore wear, and air density and temperature (which affects, among other things, the rate at which the propellant burns, itself varying from batch to batch). ENIAC was preempted upon completion in 1945 for decidedly nonroutine calculations concerning the feasibility of a hydrogen bomb. Although firing table computations could be checked against manual calculations, those for the hydrogen bomb could not; ENIAC's results could only be trusted (or distrusted) based on technical judgment, just as is true for today's most complicated calculations. Herman H. Goldstine, *The Computer from Pascal to von Neumann* (Princeton, NJ: Princeton University Press, 1972), pp. 225ff.

14. See, e.g., James H. Fetzer, "Philosophical Aspects of Program Verification," *Minds and Machines*, Vol. 1, 1991, pp. 197–216. Although individual blocks of code can be checked and verified independently of one another, interactions matter, as for any complex system, and may trigger failures at some later date.

15. Alfred V. Aho, "Software and the Future of Programming Languages," *Science*, Vol. 303, February 27, 2004, pp. 1331–1333.

16. *Report of the Defense Science Board Task Force on Defense Software* (Washington, DC: Office of the Under Secretary of Defense for Acquisition and Technology, November 2000), pp. ES-2 and 12. The classic discussion of software engineering and management remains Frederick P. Brooks, Jr., *The Mythical Man-Month* (Reading, MA: Addison-Wesley, 1975). That this book has not been superseded attests to the slow pace of advance in management of software projects.

17. As already noted, DOD attributed 40 percent of fiscal 2003 RDT&E expenditures, or about $21 billion, to software. Procurement costs for software are close to zero (once developed, the code is simply replicated for storage). But software maintenance is expensive, and mostly paid out of operations and maintenance accounts. These provide no breakdown for software. While the total cannot be estimated with any accuracy, DOD's software spending might easily be three times its expenditures for software RDT&E.

18. *Transforming Defense: National Security in the 21st Century* (Washington, DC: National Defense Panel, December 1997), p. 43.

19. Hence the title of Admiral Bill Owens with Ed Offley, *Lifting the Fog of War* (New York: Farrar, Straus and Giroux, 2000). Because warships have usually been expected to fight as part of a battle group, within which close coordination is necessary for mutual defense, many network-centric concepts originated in the Navy.

20. Much of the RMA literature is unsatisfyingly nebulous. Among the exceptions are Lawrence Freedman, *The Revolution in Strategic Affairs*, Adelphi Paper 318 (Oxford, UK: Oxford University Press for the International Institute of Strategic Studies, April 1998); and Jeremy Shapiro, "Information and War: Is It a Revolution?" *Strategic Appraisal: The Changing Role of Information in Warfare*, Zalmay M. Khalilzad and John P. White, eds. (Santa Monica, CA: RAND Project Air Force, 1999), pp. 113–153.

21. "The foundation of [network-centric operations] proceeds from a simple proposition: the whole of an integrated and networked system is far more capable than the sum of its parts." *The National Defense Strategy of the United States of America* (Washington, DC: Department of Defense, March 2005), p. 14. The proposition could hardly be in greater conflict with the preferences of the services to do things their own way.

22. Thus [Vice Admiral] Arthur K. Cebrowski and John J. Garstka, in "Network-Centric Warfare: Its Origin and Future," *U.S. Naval Institute Proceedings*, Vol. 124, No. 1, January 1998, pp. 28–35, laud Wal-Mart for "network-centric retailing." If supply chain management has been a key to Wal-Mart's growth, so has the decidedly old-fashioned strategy, unmentioned by Admiral Cebrowski (who served as the Pentagon's Director of Force Transformation from 2001 until 2005) and his coauthor,

of pushing down supplier prices through market power while paying low wages to a high-turnover in-store labor force.

23. In one study, no individual could be found in three-quarters of banks surveyed who was able to explain even a relatively simple process such as all the steps (internal to the bank) in opening a new account. Baba Prasad and Patrick T. Harker, "Examining the Contribution of Information Technology toward Productivity and Profitability in U.S. Retail Banking," Working Paper 97–09, Financial Institutions Center, Wharton School, University of Pennsylvania, Philadelphia, March 1997, pp. 23–24.

24. *Defense Science Board 2005 Summer Study on Transformation: A Progress Assessment, Volume II: Supporting Reports* (Washington, DC: Office of the Under Secretary of Defense for Acquisition, Technology and Logistics, April 2006), p. 20.

25. See, e.g., Paul A. Strassmann, *The Squandered Computer: Evaluating the Business Alignment of Information Technologies* (New Canaan, CT: Information Economics Press, 1997). While many business IT projects fail, perhaps abysmally, some firms manage large gains. Losers then imitate winners, successful innovations are "selected" and spread through the sector or, if generic, the entire economy. Thus frequent failures are not incompatible with large productivity gains over time.

26. *Realizing the Potential of C4I: Fundamental Challenges* (Washington, DC: National Academy Press, 1999), p. 40.

27. *Realizing the Potential of C4I*, p. 46.

28. "Address by General W.C. Westmoreland, Chief of Staff, United States Army, Annual Luncheon, Association of the United States Army, Washington, DC, October 14, 1969," reprinted in Paul Dickson, *The Electronic Battlefield* (Bloomington: Indiana University Press, 1976), pp. 215–223; quotation from pp. 220–221.

29. Stephen D. Biddle, "Allies, Airpower, and Modern Warfare: The Afghan Model in Afghanistan and Iraq," *International Security*, Vol. 30, No. 3, Winter 2005/06, pp. 161–176; quotation from p. 170.

30. Sean Naylor, *Not a Good Day to Die: The Untold Story of Operation Anaconda* (New York: Berkeley, 2005), p. 281.

31. Quoted in Paolo E. Coletta, *The United States Navy and Defense Unification, 1947–1953* (Newark, DE: University of Delaware Press, 1981), pp. 277–278. During Operation Anaconda, American forces had no artillery; because of poor intelligence they expected little resistance and went into the area with only a few mortars. Air support was then essential, but planning had been poor, as recounted in Benjamin S. Lambeth, *Air Power Against Terror: America's Conduct of Operation Enduring Freedom* (Santa Monica, CA: RAND, 2005), pp. 163–231.

32. The quoted phrase on JTIDS and MIDS is from Myron Hura, et al., *Interoperability: A Continuing Challenge in Coalition Air Operations* (Santa Monica, CA: RAND, 2000), p. 200.

33. "Joint Warfighting: Attacking Time-Critical Targets," GAO-02–204R, letter to the Honorable Jerry Lewis from James F. Wiggins, Director, Acquisition and Sourcing Management, U.S. General Accounting Office, November 30, 2001, p. 1.

34. Michael A. Palmer, *Command at Sea: Naval Command and Control since the Sixteenth Century* (Cambridge, MA: Harvard University Press, 2005), p. 263.

35. Richard G. Davis, "Strategic Bombardment in the Gulf War," R. Cargill Hall, ed., *Case Studies in Strategic Bombardment* (Washington, DC: U.S. Government Printing Office, 1998), pp. 527–621; quotation from p. 556.

36. Quotation from an anonymous officer interview in Michael Barzelay and Colin Campbell, *Preparing for the Future: Strategic Planning in the U.S. Air Force* (Washington, DC: Brookings, 2003), p. 69.

Chapter 9   Innovation and Learning in Peace and in War

1. New York: Putnam's, 1980, pp. 172 and 174.

2. Nathan Rosenberg, *Inside the Black Box: Technology in Economics* (New York: Cambridge University Press, 1982), pp. 120–140; Eric von Hippel, *The Sources of Innovation* (New York: Oxford University Press, 1988).

3. W.A. Brockett, G.L. Graves, M.R. Hauschildt, and J.W. Sawyer, "U.S. Navy's Marine Gas Turbines," *Naval Engineers Journal*, Vol. 78, 1966, pp. 217–240; Figure 18, p. 236. On the lessons of airline operating experience more generally, see G. Philip Sallee, *Economic Effects of Propulsion System Technology on Existing & Future Transport Aircraft*, NASA CR 134645 (Washington, DC: National Aeronautics and Space Administration, July 1974).

4. Because learning by production workers is informal, economists often call it learning-by-doing. However, *all* forms of technological learning are heavily experiential; thus "learning in production" is the more accurate term.

5. DOD would claim to accomplish something like continuous improvement through "spiral development," which adds capability in "blocks" with performance, in theory, "spiraling" upward. A valid approach for systems that can be fitted with improved sensors and electronics, spiral development has less to offer when overall design architecture largely determines ultimate capability. It can also drive up costs. A related approach put forward in earlier decades under the label "concurrent development," in essence an effort to shorten acquisition cycles by ordering tooling and beginning production before the design had been frozen, usually led instead to lengthier schedules (and higher costs) because too many engineering changes were needed to resolve problems uncovered too late during development and testing. Michael E. Brown, *Flying Blind: The Politics of the U.S. Strategic Bomber Program* (Ithaca, NY: Cornell University Press, 1992), provides many examples.

6. Franklin C. Spinney, "Defense Facts of Life," staff paper, U.S. Department of Defense, December 5, 1980, pp. 96–121.

7. Richard H. Kohn, "The Early Retirement of Gen Ronald R. Fogleman, Chief of Staff, United States Air Force," *Aerospace Power Journal*, Spring 2001, pp. 6–23, quotation from p. 14.

8. Richard R. Muller, "Close Air Support: The German, British, and American Experiences, 1918–1941," Williamson Murray and Allan R. Millett, eds., *Military Innovation in the Interwar Period* (Cambridge: Cambridge University Press, 1996), pp. 144–190; quotation from p. 189.

9. The literature on World War II strategic bombing is voluminous. Stephen L. McFarland and Wesley Phillips Newton, "The American Strategic Air Offensive against Germany in World War II," *Case Studies in Strategic Bombardment*, R. Cargill Hall, ed. (Washington, DC: U.S. Government Printing Office, 1998), pp. 183–252, includes a bibliographic essay.

10. Mark Clodfelter, "Pinpointing Devastation: American Air Campaign Planning before Pearl Harbor," *Journal of Military History*, Vol. 58, 1994, pp. 75–101. Planners assumed, incorrectly, that the structure of the German economy resembled that of the United States. They also believed the German army to be more heavily mechanized, hence dependent on gasoline, than it actually was.

11. Curtis LeMay, a group commander early in the war, recalled that "The navigators . . . . had had one ride in a B-17 before they navigated across the Atlantic[.] The bombardiers had never dropped a live bomb. They'd dropped some practice bombs over a desert on a nice white circle you could see for fifty miles . . . ." "Discussion," *Air Power and Warfare: The Proceedings of the 8th Military History Symposium, United States Air Force Academy, 18–20 October 1978*, Alfred F. Hurley and Robert C. Ehrhart, eds. (Washington, DC: U.S. Government Printing Office, 1979), p. 197. Not all the B-17s made it across the Atlantic; six or seven of every 100 were forced down by mechanical failure or else got lost, ran out of fuel, and ditched at sea or in the Arctic.

12. The British had no better information for selecting targets than the Americans, at first gave their pilots only rudimentary training in night flying, and failed to appreciate that each bomber would need its own navigator. Upon realizing that no more than one-fifth of bombs dropped at night hit within five miles of their intended target, the Royal Air Force gave up the effort to find and destroy point targets such as factories and switched to area bombing of cities (and, when its planes could not find cities, farm fields). Richard Overy writes, "So inaccurate was British bombing that German intelligence had considerable difficulty in grasping exactly what strategy the British were pursuing." *Why the Allies Won* (New York: Norton, 1996), p. 108.

13. In 1940, after several months in England observing Royal Air Force operations, Carl Spaatz returned to the United States "reaffirmed [in] his belief

in the necessity of daylight precision bombing." He "apparently had not fully realized that the British had switched to night operations because of excessive casualties suffered in daylight operations." Richard G. Davis, *Carl A. Spaatz and the Air War in Europe* (Washington, DC: Center for Air Force History, 1993), p. 53.

14. See, e.g., Wesley Frank Craven and James Lea Cate, eds., *The Army Air Forces in World War II: Volume Three, Europe: Argument to V-E Day, January 1944 to May 1945* (Washington, DC: Office of Air Force History, 1983 [reprint of 1951 University of Chicago Press edition]), which notes on p. 45 that "For the first half of 1944 . . . actual production [of German single-engine fighters] reached a monthly average of 1,581, whereas Allied intelligence estimated only 655." Poor operational intelligence, a long-standing weakness in an organization accustomed to treating staff positions such as intelligence as dumping grounds for second-rate officers, hampered the U.S. Army throughout the war.

15. Quoted on p. 37 in Richard G. Davis, "Bombing Strategy Shifts, 1944–45," *Air Power History*, Vol. 36, Winter 1989, pp. 33–45. In the late stages of the war, the United States went so far as to consider sending obsolete or worn-out ("war weary") bombers packed with explosives on unmanned flights to crash in German cities. Ronald Schaffer, "American Military Ethics in World War II: The Bombing of German Civilians," *Journal of American History*, Vol. 67, 1980, pp. 318–334. The plan was eventually abandoned.

16. Nearly 30,000 American airmen died in the European bombing campaign (including fighter pilots attached to the 8th and 15th Air Forces). As described by Davis, *Spaatz and the Air War in Europe*, pp. 371–372,

> [T]he B-17s . . . grouped in a sixty-mile-long column. . . . took a fearful beating. Bf 109s and FW 190s dived and twisted through the bombers' formation, whose gunners tried futilely to keep them at bay.
>
> [T]he American pilots . . . dared not try any but the slightest wobble of evasive action for fear of crashing into each other in formation. Once a bomber spouted smoke or flames and, laboring, fell back from the safety of the formation, the fighters finished it off . . . . The crew of a fatally injured bomber had little time to escape before hundreds of gallons of fuel or six thousand pounds of bombs exploded. If the bomber began to spin, the centrifugal force it generated trapped the crew within it. Many airmen were crushed and broken by the tail surfaces of their own bombers when the slipstream grabbed them as they exited . . . . Anyone who bailed out started a five-mile descent to ground by falling through other bomber formations, perhaps meeting grisly death on props and leading wing edges.

17. R.J. Overy, *The Air War, 1939–1945* (New York: Stein and Day, 1981), p. 78.

18. Overy, *The Air War*, p. 100.

19. Planners knew there were few if any targets of military significance in the two cities. If the United States had not already become accustomed to the fire-bombing of urban areas, would it have been so ready to drop atomic bombs on these secondary targets, or indeed on any targets? That is a major theme in Michael S. Sherry, *The Rise of American Air Power: The Creation of Armageddon* (New Haven: Yale University Press, 1987), who finds in the bombing of civilians, by then hardly questioned, a principal reason for the near-total absence of objections to the use of atomic weapons following their demonstration at White Sands.

20. During the 1991 Gulf War, oil company surveys provided the only means of relating GPS readings to observed terrain features in a number of desert areas. Those surveys sometimes proved to be off by hundreds of yards. Michael Russell Rip and David P. Lynch, "The Precision Revolution: The Navstar Global Positioning System in the Second Gulf War," *Intelligence and National Security*, Vol. 9, 1994, pp. 167–241. Errors this large, while not serious for guiding tank columns, render GPS-guided munitions useless.

21. According to reports of studies conducted on the ground by the U.S. military after the Serb withdrawal, NATO planes destroyed as few as 14 tanks rather than the 120 initially claimed, 18 armored personnel carriers rather than a claimed 220, and only 20 mortars and artillery pieces rather than 450. John Barry and Evan Thomas, "The Kosovo Cover-Up," *Newsweek*, May 15, 2000, pp. 23–26. The initial figures had been based on aircrew reports—since the early days of bombing in World War I almost universally found to be overstated—and aerial or space-based surveillance. Even if the actual totals were several times greater than reported by *Newsweek*, Serbian military capabilities would not have been seriously harmed.

22. Fighter pilots in World War II, told to "shoot up any moving target in Germany," fired "at cows in the fields, pedestrians, bicycle riders, farm carts, and any moving civilian or military vehicles." Davis, *Spaatz and the Air War in Europe*, pp. 378 and 711.

23. The "for want of a nail" notion underlying strategic bombing as propounded in the prewar Air Corps—that attacks on a few key sources of supply, such as ball bearing factories, would trigger military collapse—may have stemmed from the fact (or belief) that the springs for the variable pitch propellers of the Army's own planes came from a single plant. Robert T. Finney, *History of the Air Corps Tactical School 1920–1940* (Washington, DC: Center for Air Force History, 1992 [1955]), p. 65.

24. Thomas C. Hone, Norman Friedman, and Mark D. Mandeles, *American and British Aircraft Carrier Development, 1919–1941* (Annapolis, MD: Naval Institute Press, 1999), provides an exemplary analysis of the interplay of national strategy, technological capability, and organizational dynamics over several decades as the U.S. and British navies sought to explore and then to resolve or accommodate the many uncertainties surrounding carrier aviation.

25. Escort carriers for convoy duty and support of amphibious landings were smaller. The evolution of aircraft carriers (and battleships) can be traced in Robert Gardiner, ed., *The Eclipse of the Big Gun: The Warship 1906–45* (London: Conway Maritime Press, 1992).

26. Followed by torpedo-firing destroyers. Wayne P. Hughes Jr., *Fleet Tactics and Coastal Combat*, Second Edition (Annapolis, MD: Naval Institute Press, 2000), pp. 123–137.

27. Robert L. O'Connell, *Sacred Vessels: The Cult of the Battleship and the Rise of the U.S. Navy* (Boulder, CO: Westview, 1991), pp. 318 and 320–321. Shore bombardment continued to excite a few enthusiasts at century's end, as illustrated in John F. Lehman, Jr. and William L. Stearman, "Keep the Big Guns," *U.S. Naval Institute Proceedings*, Vol. 126, No. 1, January 2000, pp. 43–47.

28. The fuze design was eventually found to be flawed. This sorry tale is a running theme in Clay Blair, Jr., *Silent Victory: The U.S. Submarine War against Japan* (Philadelphia and New York: Lippincott, 1975). Hone, Friedman, and Mandeles, *American and British Aircraft Carrier Development*, pp. 170–171, contrast the attitudes of Navy aviators with the conservative behavior of submariners. Less than aggressive tactics, especially in the first two years of the war (during which most of the overly cautious commanders were weeded out) harmed the credibility of claims that torpedoes struck their targets but failed to explode.

29. Other suggestions included conversion to missile platforms, antisubmarine warfare, landing support, communications or tactical command centers, and supply or heavy repair ships. Lawrence C. Allin, "An Antediluvian Monstrosity: The Battleship Revisited," *Naval History: The Seventh Symposium of the U.S. Naval Academy*, William B. Cogar, ed. (Wilmington, DE: Scholarly Resources, 1988), pp. 284–292.

30. Brian Balogh, *Chain Reaction: Expert Debate and Public Participation in American Commercial Nuclear Power, 1945–1975* (Cambridge: Cambridge University Press, 1991), p. 113.

31. Light water reactors capture the heat produced by fission in ordinary or light water, as opposed to heavy water (the hydrogen in heavy water contains a neutron as well as a proton, doubling the atomic weight), one of several alternative moderators.

32. Rickover seems to have agreed to manage the Shippingport project to keep the initial design work from being abandoned after his efforts to persuade DOD and the Navy to build a nuclear-powered carrier had stalled because of high costs (an estimated 50 percent premium for advantages that were decidedly modest compared to those for nuclear propulsion in submarines). Richard G. Hewlett and Francis Duncan, *Nuclear Navy, 1946–1962* (Chicago: University of Chicago Press, 1974), pp. 196–204 and 225–254. Francis Duncan, *Rickover and the Nuclear Navy: The Discipline of Technology* (Annapolis, MD: Naval Institute Press, 1990), traces the long-running subsequent conflict between the Navy and

Secretary of Defense Robert McNamara over nuclear power for surface ships following completion of the nuclear-powered carrier Enterprise in the early 1960s.

33. Richard F. Hirsh, *Technology and Transformation in the American Electric Utility Industry* (Cambridge: Cambridge University Press, 1989), pp. 116 and 241, cites an estimate by the Department of Energy putting the value of federal subsidies over the period 1950–1980 at $37 billion, evidently for R&D and nuclear fuel only.

34. Irvin C. Bupp and Jean-Claude Derian, *Light Water: How the Nuclear Dream Dissolved* (New York: Basic, 1978), p. 74.

35. Bupp and Derian write, "We found no indication that anyone raised the . . . fundamental point of the uncertainty in making cost estimates for nuclear plants for which there was little prior construction experience. . . ." *Light Water*, p. 46.

36. Richard G. Hewlett, "Science and Engineering in the Development of Nuclear Power in the United States, *Bridge to the Future: A Centennial Celebration of the Brooklyn Bridge*, Margaret Latimer, Brooke Hindle, and Melvin Kranzberg, eds. (New York: New York Academy of Sciences, 1984), pp. 193–202.

37. Many business firms, regardless of how rigorously they scrutinize other investments, have spent large sums on information technology without making much of an effort, before the fact or after, to evaluate paybacks. See, e.g., Daniel E. Sichel, *The Computer Revolution: An Economic Perspective* (Washington, DC: Brookings, 1997), p. 92. The primary reason is the absence of normative frameworks for conceptualizing internal sources of efficiency.

38. The preceding discussion is based in part on John A. Alic, David C. Mowery, and Edward S. Rubin, *U.S. Technology and Innovation Policies: Lessons for Climate Change* (Arlington, VA: Pew Center on Global Climate Change, November 2003), pp. 39–40.

Chapter 10   Generation of Variety and Selection of Innovations

1. *Capitalism, Socialism and Democracy* (New York: HarperPerennial, 1975 [1942]), pp. 82–83.

2. Both historians and economists have offered evolutionary accounts of technological development. See, e.g., George Basalla, *The Evolution of Technology* (Cambridge: Cambridge University Press, 1988); and Richard R. Nelson and Sidney G. Winter, "Evolutionary Theorizing in Economics," *Journal of Economic Perspectives*, Vol. 16, No. 2, Spring 2002, pp. 23–46. Evolutionary perspectives are also useful in understanding how firms and industries begin, grow, and decline.

3. Ray M. Haynes, "The ATM at Age Twenty: A Productivity Paradox," *National Productivity Review*, Vol. 9, 1990, pp. 273–280; Allen N. Berger,

Anil K. Kashyap, and Joseph M. Scalise, "The Transformation of the U.S. Banking Industry: What a Long, Strange Trip It's Been," *Brookings Papers on Economic Activity, 2*, 1995, pp. 55–218.

4. Stephen A. Herzenberg, John A. Alic, and Howard Wial, *New Rules for a New Economy: Employment and Opportunity in Postindustrial America* (Ithaca, NY: Cornell University Press, 1998), pp. 85–90.

5. John Jewkes, David Sawers, and Richard Stillerman, *The Sources of Invention*, Second Edition (London: Macmillan, 1969); Raymond S. Isenson, "Project Hindsight: An Empirical Study of the Sources of Ideas Utilized in Operational Weapon Systems," *Factors in the Transfer of Technology*, William H. Gruber and Donald G. Marquis, eds. (Cambridge, MA: MIT Press, 1969), pp. 155–176.

6. Linear models, which portray innovation as following a relatively direct path from research to applications, became popular after World War II, in large part due to Vannevar Bush's famous 1945 report, *Science—The Endless Frontier: A Report to the President on a Program for Postwar Scientific Research* (Washington, DC: National Science Foundation, 1990 [July 1945]). Bush opened by stating (p. 10)

> We all know how much the new drug, penicillin, has meant to our grievously wounded men on the grim battlefronts of this war. . . . Science and the great practical genius of this Nation made this achievement possible.
>
> Some of us know the vital role which radar has played. . . . Again, it was painstaking scientific research over many years that made radar possible.

His encomium to the "painstaking scientific research" underlying radar was no more than a half-truth, as Bush well knew, in that radio and television engineering, much of it conducted on a cut-and-try basis, had contributed greatly during the 1930s. His aim was to persuade policymakers that government should continue to fund research, and do so liberally, even though the war was ending.

7. John Ziman writes, in *Reliable Knowledge: An Exploration of the Grounds for Belief in Science* (Cambridge: Cambridge University Press, 1978), pp. 107–108:

> The objectivity of scientific knowledge resides in its being a social construct . . . . We know that it has been through a thorough process of criticism by well-motivated and skillful experts . . . . The very fact that it now belongs to the consensus of the scientific community implies that it is "believed" by its most expert potential critics . . . . The objectivity of well-established science is thus comparable to that of a well-made map, drawn by a great company of surveyors who have worked over the same ground along many different routes.

8. Larry Owens, "The Counterproductive Management of Science in the Second World War: Vannevar Bush and the Office of Scientific Research and Development," *Business History Review*, Vol. 68, 1994, pp. 515–576; quotations from pp. 529 and 537.

9. Mario Bunge, *Intuition and Science* (Englewood Cliffs, NJ: Prentice-Hall, 1962), p. 68.

10. *Science and Engineering Indicators 2006* (Arlington, VA: National Science Board/National Science Foundation, 2006), Appendix Tables 4–5 and 4–9, pp. A4–8 and A4–16.

11. The inventors of the transistor sent a paper announcing their discovery to the *Physical Review*, asking the editors not to print it until DOD had completed a classification review. The invention became known to the public—and subject to scrutiny by solid-state physicists—only after the Pentagon concluded that secrecy was unnecessary. Lillian Hoddeson, "The Discovery of the Point-Contact Transistor," *Historical Studies in the Physical Sciences*, Vol. 12, 1981, pp. 41–76.

12. *Ada and Beyond: Software Policies for the Department of Defense* (Washington, DC: National Academy Press, 1997).

13. On the good reasons for such seeming blunders, see Paul Ceruzzi, "An Unforeseen Revolution: Computers and Expectations, 1935–1985," *Imagining Tomorrow: History, Technology, and the American Future*, Joseph J. Corn, ed. (Cambridge, MA: MIT Press, 1986), pp. 188–201.

14. B.R. Schlender, "Intel's Development of 386 Computer Chip Took $100 Million and Four Years of Difficult Work," *Wall Street Journal*, August 29, 1986, p. 8.

15. Ralph Katz, "How a Band of Technical Renegades Designed the Alpha Chip," *Research-Technology Management*, November–December 1993, pp. 13–20.

16. The contrasting paths taken by the responsible individuals in the two companies can be traced in Michael F. Wolff, "The Genesis of the Integrated Circuit," *IEEE Spectrum*, August 1976, pp. 45–53.

17. VHSIC channeled R&D funds to teams that combined semiconductor firms and established defense contractors. The intent was to invigorate knowledge interchange and spur applications of the most advanced chips in military electronics. Semiconductor firms were less than enthusiastic participants, VHSIC came to be dominated by large military contractors, and little in the way of innovation resulted. John A. Alic, Lewis M. Branscomb, Harvey Brooks, Ashton B. Carter, and Gerald L. Epstein, *Beyond Spinoff: Military and Commercial Technologies in a Changing World* (Boston: Harvard Business School Press, 1992), pp. 268–272.

18. Among its other R&D objectives, DOD's Integrated High-Performance Turbine Technology program, which began in the late 1980s and ran for more than a dozen years (a successor program is now in place), sought a 50 percent reduction in fuel consumption. *The Department of Defense Critical Technologies Plan*, for the Committees on Armed Services, United States Congress, March 15, 1990, pp. A-154.

19. *Defense Acquisitions: Reduced Threat Not Reflected in Antiarmor Weapon Acquisitions*, GAO/NSIAD-99–105 (Washington, DC: U.S. General Accounting Office, July 1999), p. 4.

20. Companies with fewer than 500 employees accounted for 4.6 percent of U.S. industrial R&D in 1980 and 17.6 percent in 2003 (the latest year available). *Science & Engineering Indicators—1993* (Arlington, VA: National Science Board/National Science Foundation, 1993), Appendix Table 4–31, pp. 369–370; *Science and Engineering Indicators 2006*, Appendix Table 4–19, p. A4–36. Some of the reported increase probably reflects an expanded survey sample.

21. [Admiral] William A. Owens, "Creating a U.S. Military Revolution," *The Sources of Military Change: Culture, Politics, Technology*, Theo Farrell and Terry Terriff, eds. (Boulder, CO: Lynne Rienner, 2002), pp. 205–219; quotation from p. 214.

Chapter 11   Taking Reform Seriously

1. *Task Force Report on National Security Organization* (Washington, DC: U.S. Government Printing Office, 1949), pp. 37–38, as cited in Paul Y. Hammond, *Organizing for Defense: The American Military Establishment in the Twentieth Century* (Princeton, NJ: Princeton University Press, 1961), p. 241.

2. Quotation from an anonymous interview in Michael Barzelay and Colin Campbell, *Preparing for the Future: Strategic Planning in the U.S. Air Force* (Washington, DC: Brookings, 2003), p. 148.

3. *Army Medium Trucks: Acquisition Plans Need Safeguards*, GAO/NSIAD-99–28 (Washington, DC: U.S. General Accounting Office, November 1998).

4. *Army Medium Trucks*, p. 5.

5. Jeffrey Record, *Beyond Military Reform* (Washington, DC: Pergamon-Brassey's, 1988), p. 51.

6. In the mid-1980s, GAO found that about half of the major systems it examined had been produced at inefficiently low rates. *Acquisition: Status of the Defense Acquisition Improvement Program's 33 Initiatives*, GAO/NSIAD-86-178BR (Washington, DC: U.S. General Accounting Office, September 1986), p. 21.

7. Mark Lorell and John C. Graser, *An Overview of Acquisition Reform Cost Savings Estimates* (Santa Monica, CA: RAND, 2001), p. 112.

8. *Defense Science Board 2005 Summer Study on Transformation: A Progress Assessment, Volume II: Supporting Reports* (Washington, DC: Office of the Under Secretary of Defense for Acquisition, Technology and Logistics, April 2006), p. 259.

9. Chester Paul Beach, Jr., Inquiry Officer, "Memorandum for the Secretary of the Navy," Department of the Navy, Office of the Secretary, Washington, DC, November 28, 1990, pp. 33–35.

10. Robert S. Gilmour and Eric Minkoff, "Producing a Reliable Weapons System: The Advanced Medium-Range Air-to-Air Missile (AMRAAM)," *Who Makes Public Policy? The Struggle for Control between Congress and the*

Executive, Robert S. Gilmour and Alexis A. Halley, eds. (Chatham, NJ: Chatham House, 1994), pp. 195–218; quotations from p. 207.

11. "It's Time to Protect the Pentagon from Itself," *Business Week*, July 4, 2005, p. 104.

12. Quoted in "A Concise History of the Organization of the Joint Chiefs of Staff, 1942–1979," Historical Division of the Joint Secretariat, Joint Chiefs of Staff, July 1980, p. 13, as cited on p. 172 of William J. Lynn, "The War Within: The Joint Military Structure and Its Critics," *Reorganizing America's Defense: Leadership in War and Peace*, Robert J. Art, Vincent Davis, and Samuel P. Huntington, eds. (McLean, VA: Pergamon-Brassey's, 1985), pp. 168–204. Marshall had begun advocating some form service "unification" in 1943.

13. Quoted on p. 175 in Warner R. Schilling, "The Politics of National Defense: Fiscal 1950," Warner R. Schilling, Paul Y. Hammond, and Glenn H. Snyder, *Strategy, Politics, and Defense Budgets* (New York: Columbia University Press, 1962), pp. 5–266.

14. Robert W. Komer, "Strategymaking in the Pentagon," *Reorganizing America's Defense*, pp. 207–229; quotation from p. 213.

15. Paul Y. Hammond, "Super Carriers and B-36 Bombers: Appropriations, Strategy and Politics," *American Civil-Military Decisions: A Book of Case Studies*, Harold Stein, ed. (Birmingham: University of Alabama Press, 1963), pp. 465–564; quotation from p. 485. Truman initially set the total at $15 billion; $600 million was later put aside in a separate account for nuclear weapons.

16. *Beyond Goldwater-Nichols: Defense Reform for a New Strategic Era, Phase 1 Report* (Washington, DC: Center for Strategic and International Studies, March 2004), p. 14; emphasis in original.

17. For a close look at command and control in Afghanistan, revealing a good deal of confusion among combat forces that included Army Rangers, Special Forces, and paratroopers, Navy Seals, intelligence operatives, air controllers, helicopter pilots and "pilots" of unmanned Predators, plus Afghan irregulars and their American handlers, and a contingent of Australians, see Sean Naylor, *Not a Good Day to Die: The Untold Story of Operation Anaconda* (New York: Berkeley, 2005).

18. On the underlying assumptions, see Deborah D. Avant, *Political Institutions and Military Change: Lessons from Peripheral Wars* (Ithaca, NY: Cornell University Press, 1994), pp. 49–75.

19. *The National Military Strategy of the United States of America: A Strategy for Today; A Vision for Tomorrow* (Washington, DC: Joint Chiefs of Staff, 2004), pp. 5 and 23.

20. *Quadrennial Defense Review Report* (Washington, DC: Department of Defense, February 6, 2006), p. A–4. The statement is that of Marine Corps General and JCS Chair Peter Pace.

21. See, e.g., Tables 1 and 2 in Jennifer M. Lind, "Pacifism or Passing the Buck? Testing Theories of Japanese Security Policy," *International Security*, Vol. 29, No. 1, Summer 2004, pp. 92–121.

22. *Defense Acquisitions: Assessments of Selected Major Weapon Programs*, GAO-06-391 (Washington, DC: U.S. Government Accountability Office, March 2006), p. 5.

23. For a survey of officer attitudes, see Deborah D. Avant and James H. Lebovic, "U.S. Military Responses to Post–Cold War Missions," Theo Farrell and Terry Terriff, eds., *The Sources of Military Change: Culture, Politics, Technology* (Boulder, CO: Lynne Rienner, 2002), pp. 139–160.

24. Colin Campbell, *Managing the Presidency: Carter, Reagan, and the Search for Executive Harmony* (Pittsburgh, PA: University of Pittsburgh Press, 1986), p. 156. Campbell notes that by the midpoint of Reagan's first term, "the political leadership was so entrenched in its indifference to analysis of defense issues that OMB's national security division had lost nearly a third of its examiners" (p. 187).

25. Richard H. Kohn, "The Early Retirement of Gen Ronald R. Fogleman, Chief of Staff, United States Air Force," *Aerospace Power Journal*, Spring 2001, pp. 6–23; quotation from p. 12.

26. *Report of the Quadrennial Defense Review* (Washington, DC: Department of Defense, May 1997), Section III (otherwise unpaginated).

27. According to Eric V. Larson, David T. Orletsky, and Kristin Leuschner, *Defense Planning in a Decade of Change: Lessons from the Base Force, Bottom-Up Review, and Quadrennial Defense Review* (Santa Monica, CA: RAND, 2001), the two-front imperative entered post–Cold War planning as an "afterthought" to the 1989–1990 Base Force review, only later becoming the "high canon for defense planning" (p. 13). The second QDR, released in 2001, backed away from the two-front scenario only slightly.

28. *Quadrennial Defense Review Report* (2006), pp. 4, 38, and 3, respectively. For the meaning assigned to "disruptive challenges," readers must go to *The National Military Strategy of the United States of America*, p. 4, which explains that these "may come from adversaries who develop and use breakthrough technologies to negate current U.S. advantages in key operational domains." There is no explanation of what those "break-through technologies" might be—for instance, whether some of them might be low-technology and improvised.

29. *Quadrennial Defense Review Report* (2006), p. 48.

30. *Quadrennial Defense Review Report* (2006), pp. 1–2 and 41.

31. Admiral Bill Owens with Ed Offley, *Lifting the Fog of War* (New York: Farrar, Straus and Giroux, 2000), p. 163.

32. Peter D. Feaver, *Armed Servants: Agency, Oversight, and Civil-Military Relations* (Cambridge, MA: Harvard University Press, 2003), Figure 6.1, p. 193.

33. [Report of the] *Defense Science Board Task Force on The Roles and Authorities of the Director of Defense Research and Engineering* (Washington, DC: Office of the Under Secretary of Defense for Acquisition, Technology and Logistics, October 2005), p. 2 of Task Force Cochairs' cover memorandum.

34. Christopher P. Gibson and Don M. Snider, "Civil-Military Relations and the Potential to Influence: A Look at the National Security Decision-Making Process," *Armed Forces & Society*, Vol. 25, 1999, pp. 193–218. Gibson and Snider view increases in graduate training for officers since the 1960s as in substantial part a reaction to the ability of McNamara and his subordinates to push through policies over military resistance.

35. James A. Winnefeld and Dana J. Johnson, *Joint Air Operations: Pursuit of Unity in Command and Control, 1942–1991* (Annapolis, MD: Naval Institute Press, 1993), p. 169, referring to documents released after the 1991 Gulf War.

ABOUT THE AUTHOR

John Alic lives in Avon, North Carolina and Washington, DC. Coauthor, with Lewis M. Branscomb, Harvey Brooks, Ashton B. Carter, and Gerald L. Epstein, of *Beyond Spinoff: Military and Commercial Technologies in a Changing World* (1992), and with Stephen A. Herzenberg and Howard Wial of *New Rules for a New Economy: Employment and Opportunity in Postindustrial America* (1998), he has taught and conducted research at several universities and worked for more than 15 years on the staff of the U.S. Congress's Office of Technology Assessment.

INDEX